BIOTECHNOLOGY BY OPEN LEARNING

Biotechnological Innovations in Crop Improvement

PUBLISHED ON BEHALF OF :

Open universiteit   and   **Thames Polytechnic**

Valkenburgerweg 167   Avery Hill Road
6401 DL Heerlen   Eltham, London SE9 2HB
Nederland   United Kingdom

BUTTERWORTH
HEINEMANN

Butterworth-Heinemann Ltd
Linacre House, Jordan Hill, Oxford OX2 8DP

 PART OF REED INTERNATIONAL BOOKS

OXFORD   LONDON   BOSTON
MUNICH   NEW DELHI   SINGAPORE   SYDNEY
TOKYO   TORONTO   WELLINGTON

First published 1991

British Library Cataloguing in Publication Data
A catalogue record for this book is
available from the British Library

Library of Congress Cataloguing in Publication Data
A catalogue record for this book is
available from the Library of Congress

ISBN 0 7506 1512 5

Composition by Thames Polytechnic
Printed and Bound in Great Britain by
Thomson Litho, East Kilbride, Scotland

Biotechnological Innovations in Crop Improvement

BOOKS IN THE BIOTOL SERIES

The Molecular Fabric of Cells
Infrastructure and Activities of Cells

Techniques used in Bioproduct Analysis
Analysis of Amino Acids, Proteins and Nucleic Acids
Analysis of Carbohydrates and Lipids

Principles of Cell Energetics
Energy Source for Cells
Biosynthesis and the Integration of Cell Metabolism

Genome Management in Prokaryotes
Genome Management in Eukaryotes

Crop Physiology
Crop Productivity

Functional Physiology
Cellular Interactions and Immunobiology
Defence Mechanisms

Bioprocess Technology: Modelling and Transport Phenomena
Operational Modes of Bioreactors

In vitro Cultivation of Micro-organisms
In vitro Cultivation of Plant Cells
In vitro Cultivation of Animal Cells

Bioreactor Design and Product Yield
Product Recovery in Bioprocess Technology

Techniques for Engineering Genes
Strategies for Engineering Organisms

Technological Applications of Biocatalysts
Technological Applications of Immunochemicals

Biotechnological Innovations in Health Care

Biotechnological Innovations in Crop Improvement
Biotechnological Innovations in Animal Productivity

Biotechnological Innovations in Energy and Environmental Management

Biotechnological Innovations in Chemical Synthesis

Biotechnological Innovations in Food Processing

Biotechnology Source Book: Safety, Good Practice and Regulatory Affairs

The Biotol Project

The BIOTOL team

**OPEN UNIVERSITEIT,
THE NETHERLANDS**
Dr M. C. E. van Dam-Mieras
Professor W. H. de Jeu
Professor J. de Vries

**THAMES POLYTECHNIC,
UK**
Professor B. R. Currell
Dr J. W. James
Dr C. K. Leach
Mr R. A. Patmore

This series of books has been developed through a collaboration between the Open universiteit of the Netherlands and Thames Polytechnic to provide a whole library of advanced level flexible learning materials including books, computer and video programmes. The series will be of particular value to those working in the chemical, pharmaceutical, health care, food and drinks, agriculture, and environmental, manufacturing and service industries. These industries will be increasingly faced with training problems as the use of biologically based techniques replaces or enhances chemical ones or indeed allows the development of products previously impossible.

The BIOTOL books may be studied privately, but specifically they provide a cost-effective major resource for in-house company training and are the basis for a wider range of courses (open, distance or traditional) from universities which, with practical and tutorial support, lead to recognised qualifications. There is a developing network of institutions throughout Europe to offer tutorial and practical support and courses based on BIOTOL both for those newly entering the field of biotechnology and for graduates looking for more advanced training. BIOTOL is for any one wishing to know about and use the principles and techniques of modern biotechnology whether they are technicians needing further education, new graduates wishing to extend their knowledge, mature staff faced with changing work or a new career, managers unfamiliar with the new technology or those returning to work after a career break.

Our learning texts, written in an informal and friendly style, embody the best characteristics of both open and distance learning to provide a flexible resource for individuals, training organisations, polytechnics and universities, and professional bodies. The content of each book has been carefully worked out between teachers and industry to lead students through a programme of work so that they may achieve clearly stated learning objectives. There are activities and exercises throughout the books, and self assessment questions that allow students to check their own progress and receive any necessary remedial help.

The books, within the series, are modular allowing students to select their own entry point depending on their knowledge and previous experience. These texts therefore remove the necessity for students to attend institution based lectures at specific times and places, bringing a new freedom to study their chosen subject at the time they need and a pace and place to suit them. This same freedom is highly beneficial to industry since staff can receive training without spending significant periods away from the workplace attending lectures and courses, and without altering work patterns.

Contributors

AUTHORS

Dr A. G. M. Gerats, University of Ghent, Belgium

Dr M. Haring, University of Amsterdam, The Netherlands

Dr H. Huttinga, Research Institute for Plant Protection, Wageningen, The Netherlands

Professor E. Jacobsen, Agricultural University, Wageningen, The Netherlands

Dr M. Koornneef, Agricultural University, Wageningen, The Netherlands

Dr K. J. Puite, Centre for Plant Breeding and Reproduction Research, Wageningen, The Netherlands

Dr W. J. Stiekema, Centre for Plant Breeding and Reproduction Research, Wageningen, The Netherlands

Professor P. C. Struik, Agricultural University, Wageningen, The Netherlands

Dr L. Visser, Centre for Plant Breeding and Reproduction Research, Wageningen, The Netherlands

Dr J. M. Vlak, Agricultural University, Wageningen, The Netherlands

Professor L. van Vloten-Doting, Agricultural Research Dept. of the Ministry of Agriculture, Wageningen and University of Nymegen, The Netherlands

EDITOR

Dr L. Jones, Botany School, University of Cambridge, UK

SCIENTIFIC AND COURSE ADVISORS

Dr M. C. E. van Dam-Mieras, Open universiteit, Heerlen, The Netherlands

Dr C. K. Leach, Leicester Polytechnic, Leicester, UK

ACKNOWLEDGEMENTS

Grateful thanks are extended, not only to the authors, editor and course advisors, but to all those who have contributed to the development and production of this book. They include Mr R. I. James, Dr M. de Kok, Dr G. M. Lawrence, Miss J. Skelton and Mrs M. Wyatt. The development of this BIOTOL text has been funded by COMETT, The European Community Action programme for Education and Training for Technolgy, by the Open universiteit of The Netherlands and by Thames Polytechnic.

Contents

How to use an open learning text viii
Preface ix

1 An introduction to plant biotechnology,
Professor L. van Vloten-Doting 1

2 Environmental aspects of plant biotechnology,
Professor L. van Vloten-Doting 17

3 Conventional plant breeding,
Professor E. Jacobsen 37

4 Plant tissue culture,
Professor P. C. Struik 66

5 Variation and mutant selection in plant cell and tissue culture,
Dr M. Koornneef 99

6 Somatic hybridisation,
Dr K. J. Puite 117

7 Gene mapping and gene isolation,
Dr A. G. M. Gerats and Dr M. Haring 137

8 Gene transfer and genes to be transferred,
Dr W. J. Stiekema and Dr L. Visser 183

9 Case Study: Baculoviruses - genetically engineered insecticides,
Dr J. M. Vlak 221

10 Diagnostics in plant biotechnology,
Dr H. Huttinga 243

Responses to SAQs 257
Appendices 287

How to use an open learning text

An open learning text presents to you a very carefully thought out programme of study to achieve stated learning objectives, just as a lecturer does. Rather than just listening to a lecture once, and trying to make notes at the same time, you can with a BIOTOL text study it at your own pace, go back over bits you are unsure about and study wherever you choose. Of great importance are the self assessment questions (SAQs) which challenge your understanding and progress and the responses which provide some help if you have had difficulty. These SAQs are carefully thought out to check that you are indeed achieving the set objectives and therefore are a very important part of your study. Every so often in the text you will find the symbol Π, our open door to learning, which indicates an activity for you to do. You will probably find that this participation is a great help to learning so it is important not to skip it.

Whilst you can, as a open learner, study where and when you want, do try to find a place where you can work without disturbance. Most students aim to study a certain number of hours each day or each weekend. If you decide to study for several hours at once, take short breaks of five to ten minutes regularly as it helps to maintain a higher level of overall concentration.

Before you begin a detailed reading of the text, familiarise yourself with the general layout of the material. Have a look at the contents of the various chapters and flip through the pages to get a general impression of the way the subject is dealt with. Forget the old taboo of not writing in books. There is room for your comments, notes and answers; use it and make the book your own personal study record for future revision and reference.

At intervals you will find a summary and list of objectives. The summary will emphasise the important points covered by the material that you have read and the objectives will give you a check list of the things you should then be able to achieve. There are notes in the left hand margin, to help orientate you and emphasise new and important messages.

BIOTOL will be used by universities, polytechnics and colleges as well as industrial training organisations and professional bodies. The texts will form a basis for flexible courses of all types leading to certificates, diplomas and degrees often through credit accumulation and transfer arrangements. In future there will be additional resources available including videos and computer based training programmes.

Preface

The application of biotechnology to improving crop quality and management holds enormous potential. It is widely believed that biotechnology will enable mankind to increase the yield of crops and also to improve the quality of the products derived from such crops. This is, however, only part of the story. Biotechnology offers many opportunities to develop environmentally 'friendly' plant varieties and agricultural practices. Thus replacement of polluting chemical pesticides by degradable, non-polluting bio-pesticides and the replacement of crops which have high energy input demands for harvesting and processing by less energy demanding varieties fall within the capabilities of biotechnology. It is also believed that biotechnology will enable an elaboration of the range of crop products, not only in conventional areas such as flower colour and shape in the ornamentals, but also in specialist, high value product sectors such as in pharmaceuticals. The importance of plant biotechnology and its application to crops cannot be doubted.

The purpose of this text is to enable the reader to understand and contribute to the application of biotechnology to crop improvement. There are, of course, innumerable crops to choose from. We cannot, however, fail to appreciate the burden a new student to plant biotechnology would be placed under if an attempt was made to examine all the major crops within a single text. We would either have to greatly overshoot the mark or to treat each crop so superficially that the text would have been of little practical value. We have reduced the available material to manageable proportions by careful selection of a mixture of broader discussion with detailed examination of specific examples. In this way, the reader is provided with both the broad context of crop biotechnology and the detailed, real problem issues that arise in applying biotechnological practices to crops.

The text begins by reviewing the main objectives of plant biotechnology and considers the environmental issues which arise using the new techniques for plant breeding. Conventional plant breeding approaches are also examined. The major part of the text however, focusses on the application of contemporary cultural methods and genetic techniques to produce new crop varieties, crop protectants and diagnostics for crop diseases. The text is not confined to technical matters and, from time to time, discusses wider issues and implications.

The experienced plant biotechnologists who made up the author/editor team have produced a user-friendly, easy access approach to this quite advanced area. They have written this text on the assumption that readers have some experience of basic plant physiology, *in vitro* cell cultivation, molecular biology and genetic engineering. Although these assumptions have been made, they have provided readers with a number of 'reminders'. Readers are strongly recommended to carry out the in-text activities since these have been incorporated and designed to significantly contribute to understanding the core text.

Scientific and Course Advisors: Dr M. C. E. van Dam-Mieras
Dr C. K. Leach

An introduction to plant biotechnology

1.1 Introduction   2

1.2 Subsequent development of the text   2

1.3 The objectives of the plant breeder   3

1.4 An overview of the important approaches to new strain development   4

1.5 Preliminary background to recombinant DNA technology   6

1.6 Biotechnology and plant secondary products   12

1.7 The potential and limitations of breeding strategies   13

1.8 Plant breeding and environmental issues   15

Summary and objectives   16

An introduction to plant biotechnology

1.1 Introduction

molecular and
cell biology

The importance of plants to mankind cannot be over emphasised. We are dependent upon plants for food and in many instances, for clothing, fuel, medicine and housing. It is not surprising therefore that much human activity and endeavour has been directed towards improving and producing useful crops. Conventional approaches to plant breeding and crop protection have achieved much. Nevertheless these approaches have their limitations and undesirable characteristics. The comparatively recent developments in our knowledge of the molecular and cellular mechanisms that underline the activities and functions of living systems have enabled us to develop novel methods for solving problems associated with crop production. These techniques focus on the application of molecular and cell biology to plant breeding and protection. It is these that provide the central theme for this text. The contribution of plant biotechnology is not, however, restricted simply to increasing the yield of crops by producing heavier crops or by generating devices for preventing the ravages of parasites and pathogens. The new technologies are leading to improvements in product quality (eg food value, shelf-life etc) and to changes in the way we use land. Plant biotechnology has, therefore, considerable potential for growth, enhancing the quality of life and the well being of the biosphere. The aim of this text is to explain how the exiting new technologies may be applied to plant systems.

importance of
traditional plant
breeding

We must however, make it clear from the beginning that we must not neglect the traditional approaches to plant breeding and crop husbandry. These traditional routes to improving crop production have contributed much and will continue to do so. The new techniques offer alternative routes to traditional objectives of plant breeding and production.

emphasis on
the application
of techniques

In this chapter, we set some objectives for the plant biotechnologists, particularly in relation to plant breeding. It is not the intention in this text to provide all of the underpinning science and technology associated with biotechnology. Here we are primarily concerned with the application of the techniques of biotechnology to problems in plant breeding and production. Nevertheless, we take the opportunity in this introductory chapter to remind you of some of the basic principles involved. Some emphasis is placed on plant tissue culture and recombinant DNA technology since these are central to subsequent chapters. We briefly outline how these techniques widen the gene pool available to the plant breeder and how they may be used to facilitate strain selection and provide enhanced quality assurance.

1.2 Subsequent development of the text

importance of
environmental
issues

On completion of Chapter 1, the reader should have an overview of the important technologies and issues involved in plant biotechnology. Amongst these issues is the concern which is expressed over the release of new plant strains into the environment or the use of new plant protection agents. Ultimately, the products of plant biotechnology will be used within the environment, therefore environmental concern is a major consideration, irrespective of the technology being used or the objectives set for the application of the technology. Chapter 2 provides an examination of the probable environmental consequences of plant biotechnology with a particular emphasis placed on dealing with the question, is it safe? Chapter 3 examines conventional breeding and explores the ways in which the new technologies may contribute to, and enhance the capabilities of conventional breeding programmes.

The remaining chapters are devoted to these new technologies. We can broadly divide these into two sub-areas:

- those techniques which depend upon developments in the understanding of cells and tissues;

- those techniques which depend upon developments in the understanding of molecular biology.

The two aspects are, of course, not mutually exclusive and there is considerable interaction between the two. Chapters 4, 5 and 6, concentrate on the application of cell and tissue culture procedures. The remaining chapters focus onto the manipulation of plant systems at the molecular level. In order to provide opportunity for an in depth appreciation of what is involved, in many of the chapters we have chosen to concentrate on particular examples. The experience this will give you will be readily transferable to related problems in alternative circumstances.

1.3 The objectives of the plant breeder

∏ Before we examine some of the technical issues, it is worthwhile establishing some objectives of the plant breeder. Before reading on, write down as many objectives as you can.

The overall aim of the plant breeder is to produce 'better' plants. But what do we mean by 'better'? Usually we mean that they:

- produce greater yields;

- have more tolerance to environmental factors (disease, drought etc);

- have greater food value (eg higher protein content);

- have better storage properties, longer shelf-life;

- are easier to handle (eg harvest, transport, process, cook);

- are more attractive (eg unique flower colour, clean, leaf shape).

You may well have written down additional objectives to those listed above. We will return to the objectives of the plant breeder in the chapter on conventional plant breeding (Chapter 3). There is however another important underpinning objective we must keep in mind. Breeders must achieve a commercial return for their efforts. Without this return they would soon be out of business.

desirable traits

In the next section, we outline the strategies available to the plant breeder to produce strains with desirable features (traits).

1.4 An overview of the important approaches to new strain development

From ancient times to the present day mankind has been completely dependent on plants for its food. Virtually without exception food consists of plant material or products derived from plant material such as meat, eggs and dairy products. Plants are also a major source of fibre for clothing, fuel, medicines, building materials and decoration. Plants, with their capacity to use sunlight for the conversion of carbon dioxide to sucrose, are the real primary producers on Earth. Considering the importance of plants it is understandable that long ago mankind started to select plant types with improved qualities so that these plant types better suited their needs.

1.4.1 Conventional breeding programmes

careful selection of parents

Since plant breeding pre-dates the written history we do not know at what time mankind started systematically to select the seeds of the best genotypes (eg corn with large ears or beans with big seeds), out of a population, but it must have been more than 5000 years ago. Initially, plant breeding was purely empirical. Only since the last century, with increased knowledge of genetics (remember Mendels laws!) has plant breeding evolved from an art to science. Breeders nowadays very carefully select parents for crossing programmes and calculate the best selection strategy. We will discuss this aspect of plant breeding in Chapter 3. This approach has historically been by far the most successful technique and most of the plants currently cultivated have been derived by these procedures.

1.4.2 Unconventional cross-pollination

embryo rescue

An essential step in plant breeding is combining genetic information from different parents by crossing. However only closely related species are normally crossable. Detailed knowledge of the plant sexual systems has enabled plant breeders to make crosses between less related species. For example pollination with pollen from distantly related species will sometimes be able to fertilise egg cells leading to embryo formation. When such hybrid seeds start to germinate, their development is often arrested in an early stage. In a number of cases, plant breeders have been able to cultivate such hybrid embryos on agar plates containing suitable nutrients. We will examine this technique, called embryo rescue, in later chapters. By this method a wider range of genetic material becomes available for crop improvement.

1.4.3 The application of tissue culture techniques

de-differentiate

re-differentiate

ortet

ramets

cloning

One of the main features distinguishing plants and animals is that differentiated plant cells can de-differentiate and re-differentiate giving rise to a different organ. This has for a long while found practical application in the use of cuttings from plants. Thus a single 'mother' plant (also called an ortet) can be used to produce a large number of genetically identical offspring (ramets). This is an example of cloning. More recently, this type of approach has been greatly extended. Tissue and cell culture techniques enable us to grow up plant cells as though they were single celled organisms. Thus, by taking a few cells from a desired plant strain, we can grow up many millions of identical cells. By careful manipulation of the chemical and physical environment of these, they can be converted back to their multicellular form. We can represent this process as shown in

Figure 1.1. Note however that most cloning techniques in plants do not use single cells. Indeed this is rarely successful.

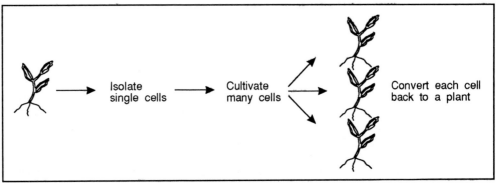

Figure 1.1 Cloning of plants using single cell cultivation. Note that with plant cells, the success rate of producing whole plants from single cells is usually quite low.

This is quite a complex process which we will discuss in greater length in later chapters. There are many potential advantages of this procedure. First we can produce many genetically identical plants. This is a particular advantage if we have a desirable parent. Secondly, while in their single cell form, we can modify the genetic content of the cells using the techniques of recombinant DNA technology and genetic engineering and, therefore we can produce completely novel plant strains. We can also produce variants from a particular plant by using mutagenic agents. We will examine this strategy in more detail in Chapter 5.

1.4.4 Genetic engineering as an approach to plant breeding

In the last decades our knowledge about cell biology, molecular biology and molecular genetics has increased to a level where this knowledge can be used to recombine genetic material very precisely. These techniques are not based on the usual sexual process in which all genes are recombined, but are based on the transfer of isolated genes by recombinant DNA technology or on the transfer of chromosomes (or complete genomes) by cell fusion. (Chapters 6 and 8) In Table 1.1 the theoretical effect of the different techniques on gene combinations is described and compared to conventional cross-pollination. Due to the limited number of traits for which the genetic code is known currently, the influence of these techniques on plant breeding is still limited. However, methods for localisation and isolation of the information for agronomically interesting traits are being developed (Chapter 7).

recombinant DNA technology

cell fusion

DNA probes

molecular markers

A spinoff from recombinant DNA technology is the use of probes and molecular markers for early selection of desired traits in breeding programmes. For example, when (part of) the genetic information for a particular trait is known, a piece of nucleic acid with this sequence can be isolated or synthesised and used as a probe. The DNA of seedlings from a breeding programme can be screened for the presence of this sequence (and thus the desired trait) by investigating whether or not this DNA can hybridise to the labelled piece of DNA. The term hybridise in this context means that two single strands of DNA have nucleotides in complimentary sequences. They will therefore, form hydrogen bonds between each other thus forming a double stranded hybrid molecule. By this method, breeding can be speeded up substantially. For example, the gene controlling the type of gluten (which confers good bread-making quality in wheat) can be detected in wheat seed. This allows selection to be made prior to planting. Without such a probe, it would be necessary to grow the wheat through several

hybridisation

gluten

Technique	Convential cross-pollination	Cell fusion	Recombinant DNA technology
Number of genes involved	Large numbers of genes re-combined	Many genes involved in recombination	Single or few genes recombined
Consequences	Many traits influenced - difficult to predict outcome	Several traits may be influenced but difficult to predict outcome	Few traits influenced - only those traits depending on single (few) gene transferable - limited application but with a largely predictable outcome
Application	Only applies to very closely related plant types	In principle can be conducted with unrelated plants - in practice it appears to work well only with related plants.	Can involve the transfer of genes between totally unrelated organisms

Table 1.1 Consequences of different techniques on gene combinations.

generations to bulk up enough seed to make flour and bake a test loaf. Unfortunately the genetic information for most traits is unknown. In such cases it may be possible to link the desired trait to a molecular marker. Selection of molecular markers enables selection for desired traits. This powerful technique is described in Chapter 7. For the next few decades the economical impact of early selection methods will probably be much larger than that of plants produced by recombinant DNA technology.

In order to be able to understand these technologies we will start with some background material.

1.5 Preliminary background to recombinant DNA technology

1.5.1 Functions of the cell

cells as functional units

All living organisms are built from cells. A cell can be considered the smallest unit of life. The actual structural organisation of different cells can be very diverse depending on the microenvironmental demands. In a unicellular organism the cell is the whole organism and must be fully self-supporting. In a multicellular organism it is part of a larger functional unit and the survival of the organism depends on a well co-ordinated co-operation between more or less specialised cells. Differences in external conditions can explain the functional variety in the appearance of the 'units of life'.

cells as chemical factories

We can easily appreciate that cellular diversity reflects an adaptation to (micro) environmental conditions. However, biochemical and molecular biological studies of a wide variety of cells have shown that, at the molecular level, different cells have quite a lot of vital strategies in common. From a molecular point of view, cells can be represented as production units that import raw materials from which they obtain building blocks for the formation of products and energy needed during the production process. Starting from these raw materials, the cells produce materials needed for their own maintenance as well as for the maintenance of other functional units within larger

| SAQ 1.1 | Make up a diagram which represents the information given in Section 1.4. To do this, re-arrange the following boxes into a representive scheme to show how the techniques described in the boxes have an input into the development of new strains. |

new plant strains	novel cross-pollination and embryo rescue
gene probes (identification of traits)	conventional cross-pollination
single/small groups of genes transferred by recombinant DNA technology	cloning by cuttings, cell culture, protoplasts
genetic engineering	large groups of genes/ complete genome transfer

organisms, and of course waste products. In other words, cells act rather like chemical factories. In order to function properly cells have internal and external sensors that register the demands from the cellular interior and from the (micro) environment. Cells can also send messenger molecules to other cells. Integration of this information directs the cellular activity.

In nature the metabolic activity of cells is mainly aimed at the maintenance and propagation of the organism and its progeny. When, however, we gain insight into the information content of cells and in the way it is expressed, we have the possibility of manipulating this information in such a way that, provided that we keep the cellular infrastructure intact and supply the cells with sufficient relevant raw materials, the cells start to make products on our demand.

In plant breeding we are interested in a number of traits such as high yields, disease and pest resistance etc. Although these traits look very diverse, they are all caused directly or indirectly by proteins.

The genetic code stored in DNA is transcribed to RNA and then translated into protein. Hence proteins are the products of expression of the genetic code and serve to mediate all the functions of the cell via their roles as enzymes catalysing the reactions of the cell's biochemical pathways.

specificity of the activities of proteins

That proteins can fulfil their cellular functions in a highly specific and efficient way is inherent in their structure. By specific, we mean that each protein has a very limited array of activities. It might, for example, only catalyse a single type of reaction. Usually enzymes work very quickly and with very few 'errors'. This high degree of specificity and efficiency also explains why, when we are looking at cells as possible production units, we will be mostly interested in their protein products. In principle when we want to 'reprogramme' organisms the experimental design is rather simple. We introduce the DNA that codes for the product into a cell in such a way that it actually becomes expressed, possibly in a specific organ, such as the fruit, flower of leaf. We then stimulate this cell to divide and regenerate into a complete plant. The approach described above sounds much simpler than it actually is. In practice many difficulties must be overcome when the experimental strategy is worked out. Furthermore one must be aware of the species specificity that occurs in nature. This means that in nature different species often differ in the signals involved in regulation of the different processes such as transcription, translation or protein export. This can have important consequences for the design of DNA to be introduced.

In the following section a general outline of the steps that have to be taken in recombinant DNA technology is given. We will examine this in much more detail in Chapters 7 and 8.

SAQ 1.2	Arrange the following terms into a sequence which relates the genetic material to the features (traits) of a plant.

Proteins   Translation

RNA   DNA

Metabolic products   Traits

Transcription

1.5.2 Recombinant DNA technology

natural and in vitro recombination of DNA

Recombination of DNA molecules is an important process in nature. The crossing-over of chromosomes during meiosis is a well known example of DNA recombination. Other examples are the processes of conjugation, transduction and transposition observed in bacteria. The really novel aspect of recombinant DNA technology is that some of the recombination steps do not take place within the cell but are carried out in a test tube. In these *in vitro* recombination steps the gene that we want to introduce into the host

vector

cell is equipped with regulatory signals and built into a so-called vector molecule, which is subsequently introduced into the host cell. After this introduction host cells that have taken up the recombinant DNA molecule and express it must be isolated from the population and regenerated. In order to be able to carry out the selection step a selectable gene, generally a gene encoding resistance against an antibiotic is included in the vector. Figure 1.2 summarises the different steps involved in recombinant DNA technology.

∏ Examine Figure 1.2. Which drug resistance gene carried by the plasmid will be functional in organism B?

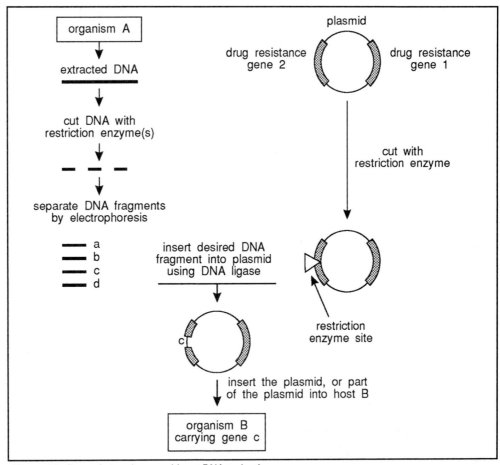

Figure 1.2 General steps in recombinant DNA technology.

You should have come to the conclusion that drug resistance gene 1 is functional. Resistance gene 2 is inactivated by the insertion of the new gene.

1.5.3 DNA or cDNA

DNA

cDNA

reverse transcriptase

introns

When we want a cell to produce a specific protein, we must introduce the DNA which codes for it into the cell. Theoretically it is possible to isolate the gene from DNA, but this approach is very tedious. Eukaryotic genomes are very large, and the chance of isolating the one gene is extremely low. Another possibility would be by *in vitro* synthesis of the DNA. But this is not realistic for a protein of normal size because it involves too many nucleotides. A third approach that is often followed in practice is the construction of a so-called copy DNA (cDNA) molecule starting from the messenger RNA (mRNA) molecule for the protein of interest. In this, a copy of the mRNA is made using the enzyme reverse transcriptase (RNA dependent DNA polymerase). In other words the nucleotide sequence in the RNA is converted into a DNA nucleotide sequence. The advantage of this approach is that within a cell manufacturing the desired protein, many copies of the mRNA are present. This facilitates the isolation of the mRNA. Furthermore, if we directly isolate the DNA containing the gene, it often contains sequences which are not translated into protein. These sequences (called introns) are removed after transcription. We can represent the flow of information as shown in Figure 1.3.

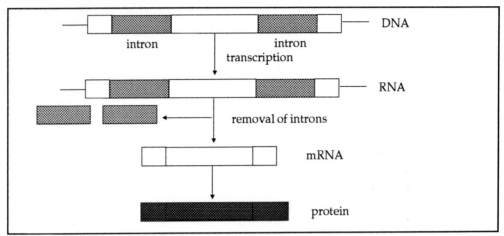

Figure 1.3 The processing of the information in DNA to produce mRNA.

Π   Use Figures 1.2 and 1.3 to explain the advantages of using cDNA over using the native gene.

There are several advantages. The main one is that if we use DNA copied against the mRNA, it does not contain any intron sequences. Thus, if we put this DNA into a new organism, we do not have to also put in a mechanism for removing the introns. The cDNA is smaller than the original gene, and is easier to pack into a vector.

The general outline for the construction of a cDNA molecule from eukaryotic mRNA is represented in Figure 1.4. There are many variations on the basic theme represented in this figure. We will deal with the processes of generating cDNA in Chapter 7.

1.5.4 *In vitro* recombination

Full details of the process involved in recombination are given in Chapters 7 and 8. We shall also learn how we can identify which vectors carry the desired DNA.

1.5.5 Transformation with recombinant DNA

use of
*Agrobacterium
tumefaciens*

When the recombinant DNA molecules have been constructed *in vitro* they must be introduced into the host cell. Several methods have been developed to introduce DNA into plant cell genomes. The most popular protocol is based on a naturally occurring system for transfer of DNA using a soil bacterium (*Agrobacterium tumefaciens*). This method, as well as some others will be discussed in Chapter 8. All these methods make use of the potential of individual plant cells to regenerate into plants.

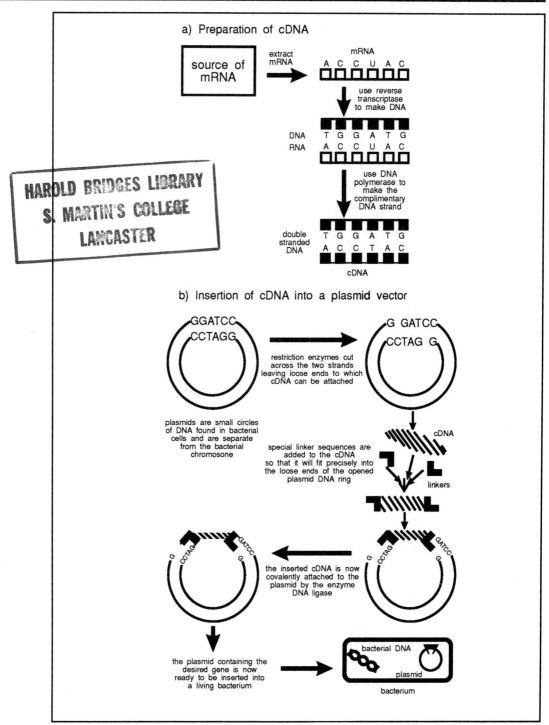

Figure 1.4 An overview of the steps in the production of cDNA and its insertion into a plasmid vector.

This type of recombinant DNA technology can only be used for traits which are based on a single gene (monogenic traits) and for which the DNA encoding this trait has been cloned. However the genetic information for only a few traits is known. Therefore, a number of methods are being developed to isolate agronomically important traits; this is a key issue of Chapter 7. Many important agricultural traits are however, based on the co-operation of several genes (polygenic traits). Transfer of uncloned larger parts of genetic information between more distantly related (uncrossable) plant species can be obtained by cell fusion (see Figure 1.5). In this, two cells are fused and the combined genetic compliments of each are re-arranged to produce hybrid nuclei containing some genetic material from both parent cells. It has proved to be possible to regenerate fusion products of some of these hybrids, however only in cases where the two parents are not too distantly related. This aspect of recombinant DNA technology is the focus of Chapter 6.

(margin notes: monogenic traits; cell fusion)

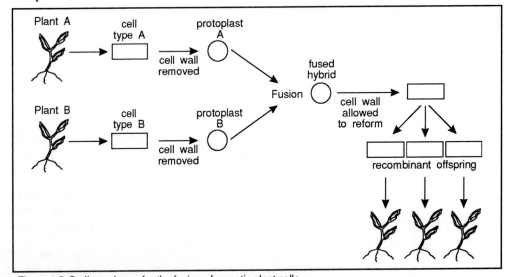

Figure 1.5 Outline scheme for the fusion of somatic plant cells.

1.6 Biotechnology and plant secondary products

Biotechnology also encompasses the industrial use of plant cells or plant organs as producers of secondary metabolites (dyes, flavours, fragrances and pharmaceuticals). The progress in this field has been rather slow partly due to the lack of knowledge of metabolic pathways in plant biochemistry - hindering the exploitation of plant cell cultures. However more important is that production of any substance by plant cells or organs in fermenters is a very costly process.

(margin note: economics of plant products produced by fermentation)

This technology is only economically feasible, for products with a very high per kilogram price such as pharmaceuticals, but not for the production of agrochemicals or for food or feed stock. Hence in this text the emphasis is put on the methods for production of transgenic plants rather than cells. The practical issues involved in the production of plant cells with high yields of important pharmaceutically active compounds are dealt with in the partner BIOTOL text '*In vitro* Cultivation of Plant Cells'. We also include aspects of crop protection because in the last few decades mankind has relied mainly on agrochemicals to protect its crops from pests and diseases. In the last few years it has become evident that many of these agrochemicals

(margin note: environmental cost of persistent agrochemicals)

are very persistent and lead to pollution of the environment endangering several species, including mankind, through pollution of sources of drinking water.

biological control of pests

The recognition of this problem has led to renewed interest in biological control of pests and diseases. However many agents showing potential for use in biological control need to be improved before they can compete in efficiency of pest control with present day methods. As an example, the approaches to improve the efficacy of an insect virus as a device for protecting crops is presented in Chapter 9.

1.7 The potential and limitations of breeding strategies

Traditionally improved varieties of crop plants have been obtained by crossing parent plants each holding part of the desired traits and selecting from the progeny those individuals showing the proper combination of traits. In this way tremendous improvements have been achieved in crop yield and quality, hence in cost of agriculture. Each day we can experience the fruits of this breeding approach at the greengrocer or supermarket. Nevertheless plant breeding by conventional crossing has its limitations. It is self evident that only traits present in related species can be combined, since crosses between distantly or unrelated species yield sterile plants or, more often, no progeny at all. Another drawback of cross-breeding is that all genes and thus traits - good and bad - of both partners are recombined. This will result in progeny plants having the desired traits combined with a number of unwanted ones. To eliminate these latter properties a time consuming programme of back-crossing is necessary. Moreover it is not always possible to get rid of undesired traits.

disease susceptibility of conventional cultivars

Thus, in spite of the impressive achievements of traditional plant breeding by crossing and selection this approach has its limitations. This is demonstrated by the fact that the most popular commercial Dutch potato cultivar Bintje (originally isolated in 1917 by a schoolteacher and hobby-breeder de Vries) with superb qualities concerning yield, cooking and chipping and for starch production, is very susceptible to all kind of plant pests like aphids (which are vectors of viruses), fungi, bacteria and nematodes. Despite extensive attempts, it has not been possible to remedy this by conventional breeding. The same holds true for Russel Burbank, one of the oldest cultivars of the US. Russel Burbank also has excellent processing qualities but lacks resistance to several diseases and nematodes.

multigenic traits

molecular breeding

As we have already seen, several techniques have been developed which permit the combination of genetic material of different organisms without making use of the sexual cycle. Traits of viruses, bacteria, fungi, plants, animals and humans can be introduced into plants, opening up a whole new range of possibilities for crop improvement. Potentially, the new approaches towards plant breeding using molecular breeding, can lead to the incorporation of any isolated gene into existing well performing plant varieties. However, molecular breeding also has its limitations. A technical problem is that at present the number of isolated genes encoding agronomically interesting traits is still very limited, but this will improve with time. A more serious problem is that traits are often determined by a series of co-operating genes. This holds especially true for the so called quantitative traits such as yield, protein content etc. Isolation and transfer of genes responsible for such a trait is very difficult or even impossible in cases where each gene separately contributes only a small part. It is also necessary to ensure that the inserted genes are correctly expressed at the right time and in the desired organ. This requires addition of regulatory sequences and, in some cases, extra DNA sequences for transport peptides.

protoplasts

cell fusion

Developments in cell biology may help to overcome this problem. Plant cells devoid of their cell wall (protoplasts) can be fused by several means. These are described in Chapter 6. In the fusion products the nuclei of the different partners also fuse. When the parent plants are (not too) distantly related fertile plants can be regenerated from these fusion products. By this method genetic material of plants, which cannot be crossed, can be recombined. Prior to fusion large parts of the genome of one of the partners can be removed or inactivated. In this way a substantial amount of uncloned genetic information (hopefully containing the information for multigenic determined traits) can be transferred, while the amount of unwanted genetic material is limited. In its advantages and disadvantages cell fusion falls between cross-breeding and molecular breeding.

SAQ 1.3

Indicate which of the following statements apply to conventional breeding using cross-pollination, which apply to recombinant DNA technology and which apply the cell fusion techniques by ticking the appropriate boxes.

	Conventional cross-pollination	Recombinant DNA technology	Cell fusion
a) This technique involves the transfer of a single or only a few genes			
b) This technique only applies to genetic recombination between closely related plants			
c) This technique cannot be applied to developing traits which are the products of many genes			
d) This technique can be used to introduce genes from bacteria			
e) The outcome of this technique is highly unpredictable			

1.7.1 A word of caution

The potential of molecular and cellular biology for advanced breeding strategies was recognised early on. At the beginning of the 1980s enthusiastic stories about tremendous and easy improvement of crop plants were repeatedly told and published. Among the 'wonderplants' predicted were: non-leguminous plants able to fix nitrogen and plants with a more efficient system of CO_2 fixation, due to incorporation of C_4 fixing genes or due to the inhibition of photorespiration etc.

Nature is much more complicated than was thought initially by molecular biologists. Nitrogen fixation has been found to be a very complex process indeed. The complexity makes the problem interesting, but frustrates the probability of transfer of this trait in the forseeable future, although considerable progress has been made in understanding the complex of nitrogen fixing genes. Wonderplants have not arisen and probably never will.

What then is reality?

Successes have been achieved in the area of breeding for virus resistance and insect resistance. New flower variants and more durable fruits have been obtained. Field trials have been (are being) performed and it is expected that within a few years the first molecularly improved varieties will be ready to be commercialized. For successful commercialisation, problems with regulation and public acceptance of transgenic plants need to be solved. We deal with these issues in Chapter 2.

What is the future?

Increasing awareness of environmental pollution by chemical pesticides is leading to a substantial reduction in permission given by governments for the use of pesticides as well as to increasing demand for 'naturally' produced food by the public. Stimulated by this development, breeding programmes, including molecular and cellular techniques, are concentrating on resistance against pests and diseases, on the use of environmental-friendly herbicides and on improved keeping quality of the edible or ornamental plant parts. Early selection using molecular probes and markers will greatly contribute to these programmes.

1.8 Plant breeding and environmental issues

The production of plants containing novel combinations of genes on a large scale pose several important problems relating to the safety of the environment and of mankind. Since these issues relate to new plant strains however they are produced, it is important that the plant breeder is aware of the issues involved. These issues are of course more acute with the development of new strains containing genes from quite unrelated organisms (ie the products of recombinant DNA technology). For this reason, we will examine the environmental aspects of plant biotechnology before we examine the techniques themselves.

Summary and objectives

In this introductory chapter we have attempted to set the scene for the remainder of this text. Now that you have completed this chapter you should be able to:

- make a list of the main objectives of the plant breeder;

- describe, in outline, the main strategies used to produce strains containing new combinations of genes;

- describe the advantages and limitations of the various techniques available to the plant breeder;

- explain why using cDNA is the chosen route for isolating genetic sequences carrying specific information.

Environmental aspects of plant biotechnology

2.1 Introduction   18

2.2 Environmental effects of transgenic plants   18

2.3 Environmental effects of transgenes and their products   21

2.4 The public debate   27

2.5 Risk analysis   33

Summary and objectives   36

Environmental aspects of plant biotechnology

2.1 Introduction

In the last decades the advances in molecular biology, molecular genetics and tissue culture have proceeded at an astonishing rate, allowing breeders to alter plants by directly modifying theirenetic make-up more quickly and more precisely than ever before. There is general agreement that this type of plant breeding can yield far reaching improvements in agricultural practice and consequently also in our environment. On the other hand there is, both in the scientific community and especially among the general public, a lot of concern about possible negative effects of such 'man-made' organisms on the environment. By 'man-made' organisms we mean plants which contain a gene or a collection of genes derived from other organisms. Such plants are

transgenic plants
described as transgenic. In the first part of this chapter the theoretical background for environmental concern of the use of transgenic plants will be presented and evaluated. Since the environmental effects of transgenic plants will be due to a combination of the characteristics of the plant and of the transgene(s) or its (their) product(s) we will first concentrate on the question: 'What determines whether or not a transgenic plant may become a nuisance?' Then we will discuss the effects of different classes of transgenes. This is followed by a section listing the most prominent questions and objections raised by public action groups. Finally, the main points to be considered when producing or using transgenic plants will be discussed.

2.2 Environmental effects of transgenic plants

Theoretically there are many different ways by which transgenic plants may have a negative effect on the environment. The main one is that the transgenic plant becomes a nuisance. We will deal with this first.

2.2.1 The progeny of the transgenic plant becomes a nuisance

weeds
In this context nuisance means an uncontrollable weed. It is estimated that weeds reduce agricultural productivity by at least 12 per cent. Thus the appearance of more weeds could seriously affect economic productivity. Besides that, weeds might invade natural ecosystems and cause substantial changes. Creating new weeds is not a trivial hazard. On the other hand plants need several characteristics to become a serious hazard.

∏ Write a list of the characteristics you might expect to determine whether or not a transgenic plant turns into an uncontrollable weed.

Since the transgenes only represent a very small part of the total genetic information present, weediness is mainly determined by the characteristics of the original plant. Especially important is whether or not this plant is already weedy. In this respect newly introduced gene(s) will only be of significance when they drastically change the survival and/or dispersal of the plant, eg the introduction of resistance against disease, herbicides or frost. Thus your answer to the question posed above should have focused onto characteristics which aid survival and dispersal of the plant.

selection may reduce weediness

Often the selection used in plant breeding is based on characteristics which decrease the potential for weediness eg selection against loss of seeds in grains, selection for large fruits with few (or no) seeds in citrus or cucumber etc.

With respect to environmental effects of transgenic variants used in field trials, cultivated plants can be classified into three categories:

Mankind-dependent plants

maize

Plants belonging to this category will quickly become extinct when no longer cultivated. An example of such a mankind dependent plant is maize. The dependence is mainly caused by the fact that maize kernels are not released from the cob and hence find no proper place for germination.

Mankind-curtailed plants

phytosanitary

For a number of plants, agricultural practice reduces dispersal. In Europe the potato can be taken as an example of such a plant. The well established procedures of removal of shoots originating from left-over tubers to control the level of soil pathogens has (together with an occasional cold winter) prevented the appearance of wild potatoes. Agricultural practices which reduce dispersal are often based on phytosanitary rules. By phytosanitary we mean good plant hygiene, predominantly by removing or destroying unwanted plant parts.

Mankind-independent plants

Plants belonging to this class are very well able to survive without mankind's intervention. Nearly all grasses belong to this category.

It is obvious that the chances that a mankind-dependent plant will turn into an uncontrollable weed is practically zero. There is one exception, namely when the newly introduced gene invalidates the properties which made this plant mankind-dependent. Such a change would be noticed during the greenhouse experiments preceding any field trials and such plants would have to be eliminated from the breeding programme. They would also probably represent an agriculturally poorer strain (eg maize plants losing their kernels).

For mankind-curtailed plants the chance that they will turn into uncontrollable weeds is very limited. In greenhouse and limited field test experiments, the behaviour of the transgenic plants can be compared to the untransformed parent. However, before releasing plant material for commercial cultivation it should be realised that seeds are often sold all over the world and that plants which are curtailed in one environment may be mankind-independent plants in another.

sterility

killer genes

The situation is completely different for mankind-independent plants. When plants from this category are found to cause unintended negative effects, it will be very difficult to eliminate them. The introduction of mechanisms for biological containment into transgenic plants of this category can be considered. Such mechanisms may be by inducing sterility or by introducing inducible killer genes, eg antisense mRNA of glutamine synthetase. We will look at mechanisms for this later. Alternatively field tests under special conditions can be used. Risk assessment for commercial release will have to concentrate on the possible effects of the introduced gene(s) in the offspring.

international seed distribution

These three categories are only relevant when discussing field tests. It is useless to distinguish mankind-curtailed plants and mankind-independent plants when debating release of seed for commerce. The reason is that the seed business is a very international trade. Crops that are efficiently curtailed in some regions may be mankind-independent and outbreeding in other places of the world.

From the points raised above, it should be self evident that data on the survival and dispersal of transgenic plants is of great importance. The key question is how does the introduction of new gene combination influence the survival and dispersal of a plant?

2.2.2 Consequences of gene transfer to wild relatives

In this, the transgene is transferred, by sexual crossing to wild relatives from the cultivated transgenic plant, and the progeny from these crossings become a nuisance.

Whether or not hybrids between cultivated plants and their wild relative are formed depends on:

- the overlap in time of the flowering periods of the cultivated plant and its wild relatives;

- the presence of wild relatives within the distance of dispersal of vital pollen;

- the appearance of successful crossings between the cultivated plant and its wild relatives.

It is evident that when transgenic plants are cultivated over extended areas and there is a chance, albeit very low, for successful crossing, hybrids will appear.

The chance that such a transgenic hybrid, once formed, will become a nuisance depends on its characteristics. For the assessment of this environmental risk the relationship between a cultivated plant and its relatives as well as the effect of the newly introduced gene(s) on this relationship, if any, need to be considered case by case. For instance the possibility of escape of pollen of transgenic oil seed rape into other *Brassica* species must be considered, since *Brassicas* interbreed freely.

2.2.3 The transfer of genes to unrelated organisms

In this, the newly introduced trait is transferred from the cultivated plant to a non-related organism, and this transgenic non-related organism turns into a nuisance.

Transfer of genetic material between non-related eukaryotes could possibly be mediated by viruses which are able to integrate into host genomes (eg DNA viruses or retroviruses). However none of the plant viruses known today are able to promote the integration of DNA into the host plant genome.

∏ Before you read on, can you write down a reason why molecular biologists have been looking very hard to find such viruses?

The answer is that viruses, which are able to integrate in and excise their genetic material out of the host genome, are known to be excellent starting material for the development of vectors. Thus you can be sure that molecular biologists are looking very hard to find such plant viruses. So far they have been without success, as is evidenced by the fact that there are currently no plant transformation protocols based solely on plant viruses.

agrobacteria and plant genes

The presence of genes under the direction of eukaryotic promoters on plasmids of the soil bacteria *Agrobacterium tumefaciens* and *A. rhizogenes*, which are inactive in bacteria and become active after transfer to plants, suggests that at least once during evolution genes from plants have been transferred to a bacterium. However the fact that even bacteria which live in a symbiotic relationship with plants (such as *Rhizobium spp.*) do not contain detectable eukaryotic genes, suggests that the transfer and establishment of plant genes in bacteria is a very rare event indeed. Although the chance of horizontal gene dispersal is very limited, it cannot be excluded. Assuming that horizontal gene transfer does take place, it may be argued for risk assessment, that the chance of the appearance of new traits in non-related organisms is only influenced by the introduction of transgenic plants when the latter carry traits new for the ecosystem.

new genes and risk assessment

In most cases the traits (eg virus resistance) or even the product of the transgene itself (eg *Bacillus thuringiensis* toxin) are not new for the ecosystem under consideration.

pleiotropic

The modification of a single gene controlling an apparently simple character often has unexpected 'pleiotropic' (multiple) effects. This is because of the close interconnection of plant metabolic pathways. A disturbance in one pathway can profoundly modify others in unexpected ways. Consequently all transgenic plants must be carefully screened for these unexpected pleiotropic effects, which may modify their ability to complete in the natural environment.

| **SAQ 2.1** | Write down at least three factors which will lead to a low chance of distribution of the introduced transgene in the wild flora during field tests. |

| **SAQ 2.2** | Make a list of the safety precautions that need to be taken during the field tests of transgenic crops which can form fertile hybrids with wild relatives. |

2.3 Environmental effects of transgenes and their products

As indicated above the environmental effect of transgenic plants will always be due to a combination of the plants and of the transgenes or their products. Especially for commercial releases the latter will be of overriding importance.

World-wide there is an interest in introducing the following traits into plants:

- resistance against plant diseases;

- resistance against plant parasites;

- resistance against herbicides;

- new flower colours;

- improvement of quality.

Let us examine these each in turn.

Π To help you remember the main points of this discussion it would be helpful to make a summary chart as you read through the following sections. We would suggest you use a full sheet of paper and use the following type of format.

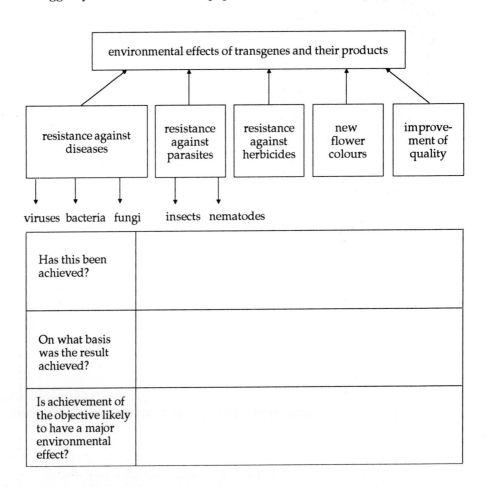

Has this been achieved?	
On what basis was the result achieved?	
Is achievement of the objective likely to have a major environmental effect?	

2.3.1 Resistance against plant diseases

Viruses

resistance based on expression of part of viral genome

At present, success has been achieved in the introduction of resistance against some viruses. The resistance is based on the introduction and expression of parts of the viral genome. In particular expression of viral coat protein protects the plant against infection with the corresponding virus. The presence of plants containing viral coat protein is not new in the ecosystem and since the amount of viral coat protein produced by transgenic plants is much lower than the amount of viral coat protein produced during a normal virus infection, no particular effects of these plants on the ecosystem are expected. It may be argued that virus resistance represents an ecological advantage which, especially in the case of mankind-independent plants, could enhance the possibility for such a plant to become an uncontrollable weed. However, evidence from 'traditional' plant breeding shows that different kinds of virus resistance have been crossed into crops and not caused any environmental problems.

Bacteria

bacterial
resistance so
far not achieved

As yet no transgenic bacteria resistant plants have been reported. It is well known that some small proteins, like thionins, apideacines and crecropines, have bactericidal properties. Attempts are being made to transfer and express genes encoding such proteins in plants. If this approach is successful the effect of these rather small and stable proteins on bacteria (both pathogenic and non-pathogenic) coming into contact with these proteins needs to be assessed. Depending on the type of transgenic plant (mankind-independent or not, fertile or sterile) these tests need to be done in greenhouses, on (small) field plots under conditions preventing outbreeding, or without any precautions.

thionins

apideacines

crecropines

∏ Can you explain why enhanced competitiveness of bacteria resistant plants may or may not be a problem?

In some cases it can be argued, as above, that traits have previously been introduced by cross-breeding and not caused environmental problems. In other cases there will be no direct parallel and 'forecasting' of effects may be difficult.

Fungi

glucanases

chitinases

No transgenic fungus resistant plants have yet been reported. At present there is quite a lot of interest in the introduction of particular genes especially glucanases and chitinases, in the hope that these enzymes will inhibit fungal growth by hydrolysing the fungal hyphal walls. These enzymes are often abundantly present in plants infected with pathogens inducing local necrosis. Hence it can be argued that the presence of plants with high activities of these enzymes is not an ecological novelty, and that the effect, if any, on the ecosystem will be limited.

2.3.2 Resistance against plant parasites

Insects

specificity of
plant-produced
toxins

broader impact
of proteinase
inhibitors

proteinase
inhibitors

Large scale use of transgenic insect resistant plants as well as the introduction of insect resistant genes into wild species via cross-breeding may influence the insect population. In its turn, a reduction of voracious insects may have an effect on wild plant species which are regulated by insect feeding. Thus, this is quite a complex matter which may have a significant environmental impact. We must, however, point out that plant produced toxins will only affect insects feeding on these plants and thus will be much more specific than the presently used chemical insecticides. On the other hand, these plant-produced toxins will probably be present during the complete (main part of) life cycle of the plant enhancing the chance of exposure. How large or small the effect of insect resistance encoding transgenes will be, will mainly depend on the (lack of) specificity of the gene product. For instance *Bacillus thuringiensis* toxins have a very high specificity. The effect will thus be limited to only a few insect species. Proteinase inhibitors may affect the ecosystem more, because they are deleterious to a broad spectrum of insects and in some cases also mammals. For risk assessment this means that interactions of transgenic plants expressing proteinase inhibitors with relevant insects and mammals need to be studied before release. It is also important to consider the effect on man. If the toxin is in a plant part normally eaten without cooking, such as in a salad crop, this could have severe clinical consequences. *Brassicas* such as cabbage and cauliflowers are frequently used fresh in salads.

Nematodes

specificity of
natural
nematocides

As yet, no transgenic nematode resistant plants have been reported. Nematodes may also be sensitive to *B. thuringiensis* toxins or proteinase inhibitors. If so, the same environmental questions discussed above have to be answered before release. Alternatively nematode resistant genes from, for example, tomato may be transferred to other crops. In this case no disturbing environmental side effects are expected, because these gene products are known to be very specific.

2.3.3 Resistance against herbicides

Herbicide resistance can be due to the introduction of a modified version of the herbicide target enzyme, over-production of the target enzyme or the introduction of enzymes able to detoxify the herbicide. It is often assumed that introduction of herbicide resistance is associated with increased usage of herbicides and thus pollution. But is this true? The answer is quite complex. Herbicides belong chemically to very different classes. Some herbicides are very persistent and pollute soil and water. However, other herbicides seem to be environmentally safe, because, in the soil, they are quickly degraded into harmless components. Introduction of resistance against this type of herbicide will lead to a shift in usage of environmentally harmful to environmentally harmless herbicides. It must be borne in mind, however that introduction of herbicide resistant genes in mankind-independent crops or by crossing into weedy wild relatives of crops, may render a particular herbicide useless. In cases where the herbicide is causing environmental problems (genes against these herbicides should not have been used in the first place), the end of its usefulness may not be a great loss or may be even beneficial to the environment. However, when dispersal of genes leads to the invalidation of one of the environmentally safe herbicides, this may be considered a loss. This is not because of the appearance of herbicide resistant plants, for in the absence of herbicide spraying they will not have an ecological advantage. The real problem is that at present there is only a limited number of these environmentally harmless herbicides and we cannot afford to diminish the use of any of them.

environmentally
harmless
herbicides

transfer of
herbicide
resistance into
weeds

The dispersal of herbicide resistant genes is dependent on:

- the survival and dispersion of the transgenic crop;

- the chance of hybridisation between the transgenic crop plants and its wild relatives.

toxicity of
herbicide
derivatives

When the herbicide resistance is based on detoxification of the herbicide, the toxicity of the herbicide derivatives for consumers need to be assessed.

2.3.4 New flower colours

inhibition of
gene
expression

new genes

In principle, two procedures are available for broadening of colour variegation. In the first inhibition of expression of genes involved in colour synthesis is used. This type of transgenic plant is comparable to deletion or dominant suppressor mutants and as such no special effects on the environment are expected. In the second approach alien genes (eg maize genes in petunia) including enzymes involved in colour synthesis are introduced and expressed in transgenic plants, leading to new products. On first sight new flower colours look absolutely innocent.

Π   Write down a reason why changing the colour of a flower may have a significant environmental effect? To answer this you will need to consider the purpose of the colouration of flowers. Write down your answer before reading on.

You may realise that due to the new colour flower other insects may visit this plant. This in turn may result in new types of cross-pollination and appearance of new hybrids. A profound knowledge of the presence of related wild species is required to estimate whether or not such hybrids may arise and/or represent an environmental risk.

2.3.5 Improvement of quality

Several crops are grown for their content of oils, proteins, pharmaceuticals, etc. Higher content of most 'usual' products will cause no significant environmental effects.

∏ On what experience is this statement based?

consequences of increased production of pharmaceuticals

It is quite evident that in the past breeders have produced plants with significantly higher levels of edible oil, proteins, etc. All these crops have been cultivated on a large scale for decades without adverse effects. Nevertheless, for some substances (especially pharmaceuticals), a concentration effect on unintended consumers may be expected and needs to be considered and/or analysed. The situation is different when plants are used to produce 'new' products. This will require a profound consideration of all environmental aspects.

Often the new gene product(s) will only be present in part of the plants. If so, only the effect of the new gene products on the (unintended) consumers of these parts is relevant. It is self-evident that in all cases where the new gene products warrant analysis, this analysis will have to be completed before a (large scale) release of the transgenic plants.

stable products and the food web

In cases where the transgene(s) is (are) responsible for the production of very stable products, not only the effect of this (these) product(s) on the primary consumers but also on the rest of the food web need(s) to be considered. In cases where the transgene leads to the formation of products which are found in plant exudates the effect of the exudates on (useful) organisms (eg bees) needs to be assayed.

In many cases the goal is to modify the composition of the seed. Examples are:

- to improve the nutritional quality of plant proteins;

- the introduction of modified proteins with enhanced lysine and methionine levels;

- the modification of the fatty acid composition of oilseeds to enhance their nutritional and processing properties (eg increase in level of mono- or poly-unsaturated fatty acids);

- the introduction of short chain fatty acids useful in industrial processes.

2.3.6 Conclusions concerning the environmental effects of transgenes and their products.

exotic plants

Based on the foregoing paragraphs it is evident that every introduction of a transgenic plant represents some risk. However, the same holds true for the introduction of plants from other continents (exotic) and for the introduction of improved cultivars obtained by cross-breeding of cultivated plants with wild species.

Comparing exotic and improved cultivars from breeding programmes with transgenic plants, the number of genes new to the ecosystem in transgenic plants in drastically less than with the exotic and improved cultivars. This simultaneously enhances the predictability of the effects of the introduction of the new genes on the environment and thus their safety! Predictions are of course no proof. Therefore, it is necessary that field tests - with adequate precautions if necessary - are performed. The enhanced predictability and safety of most transgenic plants is in contrast with the public perception of these plants (Figure 2.1).

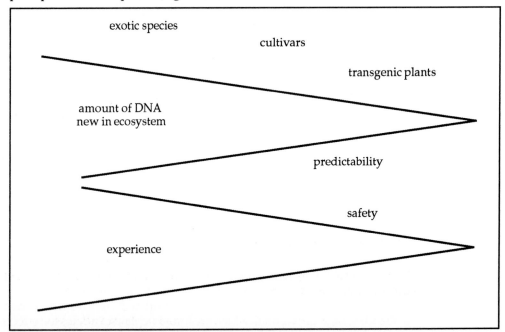

Figure 2.1 Relationship between the amount of DNA new to the ecosystem and the predictable safety. (NB public perception is often that transgenic plants are more dangerous than exotic species yet the introduction of exotic species introduces more DNA that is new to an ecosystem, than that introduced by transgenic plants).

Due to lack of proper information, or the distribution of misinformation, transgenic plants are often thought to be dangerous and emotionally associated with control of food sources by one or a few multi-nationals. Since transgenic plants resistant against pathogens, parasites and environmentally safe herbicides may contribute significantly to a decrease of environmental pollution, it is very important to avoid or correct an unjust negative image of plants improved by genetic manipulation.

To this end a large scale distribution of information to the general public, as well as an internationally accepted system of rules for field tests and release are required.

| **SAQ 2.3** | Answer true or false to each of the following statements: |

1) The introduction of genes coding for viral coat proteins often provides protection against infection by viruses.

2) It is unlikely that production of viral coat proteins by plants will have a major impact on the environment.

3) It is difficult to predict the environmental consequences of introducing the genes coding for the bactericidal proteins such as thiones into plants.

4) The introduction of chitinase genes into plants will be ecologically disasterous because this will hydrolyse chitin present in the environment.

5) Introduction of proteinase inhibitors or genes into plants will have little environmental impact other than restricting the spread of insects which feed on such transgenic plants.

6) The introduction of herbicide resistance genes into cultivated plants will invariably be environmentally advantageous.

2.4 The public debate

The questions or objections raised by public action groups about the use of transgenic plants can be classified as follows:

- possible negative effects of transgenic plants on the wild flora;

- possible negative effects of transgenic plants on the fauna including insects;

- possible increases in environmental pollution by the increased use of herbicides due to herbicide resistant crop plants;

- possible negative effects of transgenic food plants on human consumers;

- possible domination of the seed market by (one) multi-national(s) companies and the fear of genetic erosion (ie loss of genetic diversity);

- monopolisation of germ plasm due to changes from breeder's rights to patent law;

- possible negative effects on the third world.

The last items do not really belong in a chapter on environmental aspects of biotechnology. However, in public discussions on the safety of genetic engineering these points are nearly always raised. Therefore some comments on these points are included.

Again you might find it useful to draw out a summary diagram to help you remember the arguments we discuss in the rest of this section. We would suggest the following format:

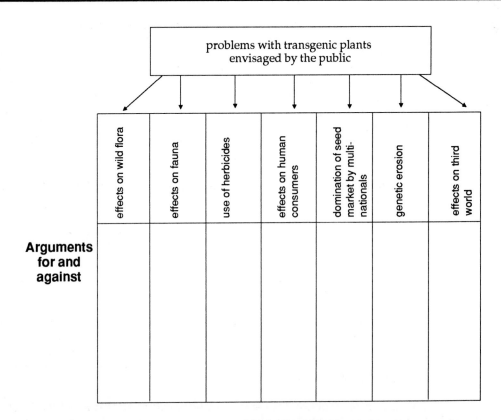

One of the main problems encountered in public discussions on safety of genetically engineered organisms in the environment is that the questions posed are very broad, while at the moment there are no generally applicable answers. Sometimes a question is best answered by presenting two or three examples showing the different salient points. This is also reflected in the system for regulation which is based on a case by case and set by step procedure.

2.4.1 Possible negative effects of transgenic plants on the wild flora

Perhaps the single most commonly voiced concern about the use of transgenic plants is that the transgenic plants or their offspring (including hybrids) may become aggressive weeds invading natural habitats and out-competing wild plants.

As we have already explained, the chance of a transgenic crop plant or its offspring becoming a weed depends to a large extent on the known properties of the plant itself and only to a minor extent on the transgene.

make a
distinction
between crop
type

In public discussions about this point one has to realise that knowledge about (crop) plant behaviour - including reproduction, survival and outbreeding - is very low in the majority of the public especially amongst inhabitants of large cities. Farmers with their accumulated knowledge of potato growing never see transgenic potatoes as potential weeds, but do realise potential problems with rape seed, or grasses.

Π   Make two lists of cultivated plants. In one make a list of those plants which are capable of becoming weeds. In the other make a list of those which are unlikely to become weeds.

To help you do this think of those cultivated plants which are capable of self-propagation (either vegetatively or by seeds) and can survive climate changes throughout the year. These could become weeds. The other list will include those plants that are, for example killed by climatic stress (eg frost in winter, drought etc) and have no mechanism for surviving such periods. Also important in placing crops in each of the lists is to consider if the plant is capable of rapid propagation (eg it produces a lot of seed) or is easily destroyed by cutting, hoeing etc. The list you generate will differ from area to area because of climatic differences. You can use your list to indicate those transgenic plants which are unlikely to become nuisance weeds, while explaining why others have the potential to become weeds.

2.4.2 Possible negative effects of transgenic plants on the fauna

extinction of insect species caused by transgenic plants appears unlikely

It is often feared that transgenic plants that are toxic for (some) animal species may lead to extinction of these species. As the products encoded by the transgenes are proteins, only species which forage on these plants will come into contact with the product of the transgene (an exception to this rule may be envisaged when the transgene encodes enzymes responsible for the synthesis of exudates). Cultivation of transgenic crop plants that are toxic to a wide array of consumers (for example by producing proteinase inhibitors) may have a negative effect on the population density of some species namely on species which are restricted to these particular crop plants. It is self-evident that the effect will be much less on insect species that can forage on a variety of plants. In general the effect of transgenic insect resistant plants will be similar to the effect of cultivation of some of the present day resistant varieties, and much less than chemical crop protection. Thus extinction of insect species due to cultivation of transgene crop plants seems to be unlikely.

when transgenes may lead to species extinction

The situation may be different when a transgene encoding a broad spectrum toxin becomes dispersed in wild relatives of the crop plant. In this case it is possible that the new trait may represent such an ecological advantage that the transgenic hybrids outcompete sensitive plant species and possibly also unrelated plants, thus limiting the food source for some animal species. As indicated before, both hybridisation with the wild flora, as well as anticipated effects of the product encoded by the transgene are points which need to be considered in risk assessments before any environmental release is permitted.

2.4.3 Possible increases in environmental pollution by the increased use of herbicides due to herbicide resistant crop plants

anti-herbicide lobby

One of the often heard criticisms against plant biotechnology is that the use of herbicide resistant transgenic crop plants will lead to an increase in the use of herbicides and thus to environmental pollution. The anti-herbicides attitude is fuelled by the feeling that agrochemicals have damaged agriculture and the environment.

not all herbicides are damaging

atrazines

phosphonitricine

There is no doubt that the availability of seeds of herbicide resistant crops will not lead to a reduction of the use of some herbicides. In all probability, the introduction of herbicide resistance in selected crops will lead to an increase in herbicide use. We have learnt however that herbicides belong to very different chemical classes. Some of them, such as atrazines, are very persistent and do have negative effects on the environment. Others, such as phosphonitricine, are rapidly degraded in the soil and are not known to have any adverse effect on the environment.

regulatory re-inforcement

Thus, by making the right choice, plant biotechnology can help to change the type of herbicides used from environment-polluting to environment-neutral. The trend can (will) be enforced by government actions forbidding the use of polluting herbicides.

2.4.4 Possible negative effect of transgenic food plants on human consumers

Partly due to an increasing awareness of the quality of food, questions about the safety of transgenic plants and their products for human consumption are getting more and more attention. Whether or not transgenic plants will be safe for human consumption depends again on the combination of the crop plant and the transgene but also on the condition in which the plant products are consumed (raw, cooked, processed etc).

In nearly all cases, transgenic plants are obtained from plants with a long history in food consumption. Usually the transgenes encodes for:

- substances which are traditionally part of the food parcel (eg viral coat proteins);

- have been tested for their safety on consumers including mankind (eg *Bacillus thuringiensis* toxin);

- are known to be inactivated by cooking or processing (eg proteinase inhibitors, but remember, many vegetables are often eaten raw);

- can be assayed for their effect on consumers.

It is relevant to point out that all products made from transgenic plants will be subjected to the normal rules about consumer safety. It appears therefore that there is little prospect of properly tested transgenic food plants proving to be dangerous to human consumers.

2.4.5 Possible domination of the seed market by (one) multi-national(s), and the fear of genetic erosion.

changes in companies involved in plant breeding

The wish to use recombinant DNA technology in plant breeding has among other things led to drastic changes in the companies involved. The investment required in personnel and equipment is often beyond the resources of the rather small (independent) breeding companies. Due to uncertainties in the patent situation, regulations and public acceptance the returns are difficult to calculate. Nevertheless, most breeding companies are convinced of the necessity of adopting this new technology to maintain their future competitive positions.

merger of plant breeding with other operations

Some breeding companies have formed joint ventures to exploit biotechnology in their common interests. Other companies have been taken over by, or merged with agro or petrochemical or pharmaceutical multi-nationals. In their turn, these multi-nationals, often already having large biotechnology departments are interested in breeding companies because of their involvement in the herbicide or pesticide market. Often they are also concerned in the long term perspectives of plants as source of raw materials to substitute for fossil energy, or chemical synthesis. At the moment it is unclear whether or not this is a lasting tendency.

tendency towards monopolies

The drastic reduction in the number of breeding companies has spawned the fear that in the future the breeding and seed market of some crops will be controlled by a limited number (one?) of companies. Apart from the fact that monopoly positions are unwanted in view of price formulation, there is the danger that the genetic variation in the breeding lines will become rather limited (genetic erosion). This may lead to cultivars sharing a large portion of their germplasm.

∏ Make a list of the risks inherent in having large areas cultivated with (nearly) identical cultivars. (Think about the consequences of changes in climate or the introduction of a virulent disease or parasite).

If only a few cultivars are in production, the occurrence of new virulent strains of disease or a persistant shift in climate, could lead to a situation in which all of these cultivars were inappropriate. It is therefore desirable to maintain a wide diversity of cultivars. Nevertheless there are considerable pressures that potentially reduce the likelihood of maintaining high numbers of cultivars. The trend is for fewer and fewer companies to be involved.

Due to the capital intensiveness of modern plant breeding, scale up of the size of breeding companies is inevitable. Safe-guards against monopoly positions can be found in anti-trust laws, while strong public breeding institutes and gene banks are necessary to protect the genetic variation.

2.4.6 Monopolisation of germ plasm due to changes from breeders' rights to patent law

Due to several facts such as:

- the cost of cloning of genes;

- the involvement of companies with other backgrounds in plant breeding;

- the imperfection of the present system of propriety protection of new cultivars (the breeder's rights);

- there has been a movement, started by biotechnology companies, to extend patent law to transgenic plants.

patent rights This movement has met with a lot of opposition. This is largely due to the fact that the first patent claims made were very wide, and it looked as if by introducing a cloned gene into a cultivar the total germplasm would fall under the patent. This is not true. Although the situation is not yet stabilised there seems to be a consensus that processes and genes can be patented when the normal requirements of the patent laws are met. We do not intend to deal with the patent law in any detail here - but the main requirements to gain patent rights are that the product must have the following features:

- novelty;

- non-obvious (ie need for creative thinking);

- usefulness;

- reproducibility.

updated breeders' rights In return for making a full public disclosure of the 'invention' - the inventor is granted monopoly rights in using the invention. One can understand why companies would prefer to protect their 'invention' by gaining patent cover. The system is not yet satisfactory and it appears likely that cultivars containing patented genes will probably, for the time being, be protected via a system of updated breeders' rights. The clarification of the legal points has diminished but certainly not abolished the opposition against patenting of biotechnologically produced plant strains. It will depend on how the patent laws are interpreted as to whether or not there will in the future be a just reward for breeders and 'gene jugglers'!

| SAQ 2.4 | Explain why it is essential that breeders do receive a fair amount of the revenues for new cultivars. |

2.4.7 Possible negative effects on the third world

There is no doubt that biotechnology in general can be a large benefit to the third world. However, for economical reasons these countries are (will) mainly be confronted with the negative side.

developed world may take third world markets

The most important negative effect that biotechnology may have on third world agriculture could be the production of traditional third world commodities or substitutes thereof such as cocoa-butter, quinine, isoglucose, coconut oil, etc, by the developed world.

decrease in traditional markets for third world products

Although it is unlikely that transgenic plants will be constructed in the near future which produce these substances, it is possible to isolate and produce enzymes which can convert 'cheap' fats to cocoa butter or corn starch to high fructose syrups or to isolate cell lines expressing quinine or other pharmaceuticals. These developments are bound to continue and will lead to a decrease in some of the traditional third world markets. To make sure that the third world is not left out of the potential benefit, international networks for improvement of third world crops such as rice and cassava have been set up. More effort in this area is clearly needed. The widespread application of recombinant DNA technology to third world crops could be severely hampered by restrictive patents and monopolisation of the technology by western based companies interested only in major crops such as maize, wheat, soya, potatoes and sugar beet.

SAQ 2.5

Indicate which of the following statements are true or false.

1) It would be environmentally sound to develop resistance to the atrazine herbicides in cultivated plants.

2) Transgenic marrows carrying the gene coding for resistance to the antibiotic kanamycin are unlikely to become nuisance weeds if cultivated in northern Europe (eg Germany, Netherlands, the UK).

3) The introduction of broad spectrum proteinase inhibitor genes into carrots to protect them against insect infestations presents no risk to humans.

4) The expense of developing transgenic plants will tend to lead to genetic erosion.

5) The development of cultivars of tropical crop plants which are capable of growing in temperate climates will benefit all of mankind because it will enable the crop to be grown in areas which operate more efficient farming and distribution practices.

2.5 Risk analysis

At present the laws, or rules for environmental release of transgenic organisms vary between countries.

But whatever the rules are, it will always be necessary to submit some kind of risk analysis when applying for permission to do a field experiment or for environmental release of transgenic organisms.

Hopefully in the near future the rules will become harmonised under the EC directives.

Below a short version of a risk analysis is presented as an example. In this risk analysis R* indicates that this statement needs to be corroborated by a literature reference or a written statement by an expert in this field. Use it to generate a list of information that is required in carrying out a risk analysis.

2.5.1 Risk analysis for a field test of transgenic potatoes in the Netherlands

Characteristics of the host

1) Description of plants used

Two varieties of potato (*Solanum tuberosum*) are used: the male sterile Bintje and the male fertile Disiree. Both genotypes have been described in the variety list (R*)

2) Potential harmfulness of the transgenic plants

feral   Potato is completely harmless for other organisms. Potatoes have been cultivated for several centuries and there are no reports of escaped (feral) potato plants surviving in Europe (R*). Feral in this context means domesticated but returned to the wild. It is not expected that incorporation of genes encoding glucuronidase resistance against the antibiotic kanamycin or the plant virus tobacco rattle virus will affect the survival of potato seeds or tubers. (R*)

3) Growth and survival

Potato is grown from seed potatoes as an annual crop. Left over tubers can sprout in the next year. In normal agricultural practices such potato plants are removed or killed to reduce the problem of soil born diseases. Potato tubers and seed are frost sensitive. Cold winters (below -5°C) completely wipe out all left-over potatoes. Desiree is a male fertile variety. Fertilisation of other potato plants by transgenic Desiree pollen will not result in a gene flow, because - as argued above, the resulting seed will not be able to establish itself in the Netherlands. In the Northern part of Europe two wild Solanum species are found, *S. dulcamara* and *S. nigrum*. Cross-breeding between these species and *S. tuberosum* has been tried extensively and is found to be impossible (R*).

4) Genetic stability

Potato is a vegetatively grown crop, showing genetic stability. There are no reports on horizontal distribution of chromosomal DNA from potato to any other species.

Characteristics of the inserts

1) Donor

The inserts have been transferred to *S. tuberosum* using *Agrobacterium tumefaciens* carrying the plasmids described (R*). The plants are free from the donor strain and only contain the DNA present between the border sequences. What is being described here is that the tumourogenic genes of Agrobacteria are not transferred into the plants - only the genes of interest together with two border regions are incorporated. (We will examine the details of this in a later chapter). The inserts consist of border sequences encompassing the glucuronidase promoters (R*); the coat protein gene for tobacco rattle virus and the kanamycin resistance gene, both driven by well known constitutive promoters (R*).

Both constructs contain very few redundant DNA sequences, insufficient to encode any unknown protein.

2) The protein encoded by the transgenes.

The proteins encoded by the transgene are all present in nature, known to be non-toxic (R*), and will, in the absence of selection pressure (kanamycin or tobacco rattle virus) not confer an ecological advantage on the transgenic plants.

Characteristics of the transgenic plants

1) Construction

The construction of the vector and the transfer to *S. tuberosum* and the characterisation of the transgenic plants are all described (R*).

Description of the field trial

1) Aim

Evaluation of the performance of transgenic plants under field conditions.

2) Location

The plants will be grown on a field plot (located R*). During the growing season the field will be inspected weekly to record the development of the plants. At the end of the growing season tubers will be harvested. Tubers will be stored and analysed at the Institute. Tubers will not be entered in the market for consumption or processing. Remaining plant material will be destroyed. Any potatoes appearing in the field in the following year will be destroyed.

3) Emergency measures

Potatoes are herbicide sensitive. In cases of unexpected negative properties of the transgenic plants, the authorities will be informed and the plants destroyed.

4) New niche

Since the potato cannot establish itself in the wild flora and the newly introduced genes will not increase its winterhardiness there is no new niche possible for these transgenic potatoes.

SAQ 2.6	It is proposed to produce a grass which is more resistant to biodegradation to use as a packing and insulation material. To do this, it is proposed to produce a transgenic grass which produces a proteinase inhibitor. Use the example given above to help you to decide what the main risks would be in the development of such a strain. Decide whether or not the project should proceed.

Summary and objectives

This chapter has examined the major environmental concerns that are expressed about developing and cultivating transgenic plants. Now that you have completed this chapter you should be able to:

- describe the different risks that arise from transgenes in mankind-dependent and mankind-independent plants;

- explain what features of plants restrict the transfer of transgenes to wild flowers during field tests;

- list precautions that should be taken when field testing transgenic crops;

- give reasoned arguments for the likely environmental risks that arise from producing transgenic disease and parasite resistant plants, changing the flower colours of plants, changing the chemical composition of plants and from introducing herbicide resistance genes;

- list the main factors which need to be considered in carrying out a risk analysis for a field test of transgenic plants;

- apply knowledge of the main factors which need to be considered in carrying out a risk analysis for a field trial to evaluate proposals for field trials.

Conventional plant breeding

3.1 Introduction   38

3.2 Plant breeding is a multi-disciplinary activity   38

3.3 The objectives of the plant breeder revisited   39

3.4 Some important terms in plant breeding   40

3.5 The strategy of the conventional plant breeder   47

3.6 Variation heritability   48

3.7 Techniques for expanding genetic variation available to the plant
    breeder   52

3.8 Mode of reproduction and its genetic consequences   58

3.9 Plant biotechnology and conventional breeding   61

Summary and objectives   64

Conventional plant breeding

3.1 Introduction

In Chapter 1, we indicated that conventional approaches to breeding have proven to be highly successful. In fact, the majority of the plants currently cultivated are the products of these conventional approaches. In a text on biotechnology and its application to plant breeding, it may seem a little strange to begin by discussing the conventional approaches. It is however important to realise that the new and conventional approaches do not operate in total isolation from each other.

We should realise that:

- conventional plant breeding strategies are, in many areas, potent competitors of the new technologies;

- there are many opportunities to integrate conventional and biotechnological approaches to achieve the desired end product. For example biotechnology, through the development of gene probes, may facilitate selection of desirable progeny from conventional crossing;

- conventional breeding offers greater opportunities for large scale gene mixing albeit between closely related plants while the new techniques usually involve more restricted gene transfer. The two types of approach therefore provide complementary strategies.

It is therefore vitally important that biotechnologists entering the field of plant breeding understand the techniques, strategies and language of the conventional plant breeder. In this chapter, we provide a description of the essential elements of conventional plant breeding and introduce the terms that are commonly employed. Towards the end of the chapter we explore the ways in which biotechnology may aid conventional plant breeding programmes.

3.2 Plant breeding is a multi-disciplinary activity

Plant breeding concerns all human activities directed to the production of plants with an improved genetic constitution to meet human needs in a better way.

factors which influence the objectives of the breeder

Let us develop a scenario of the concerns of the plant breeder. The breeder delivers varieties of plants which are source materials for the grower in agriculture and horticulture. The harvested products can be used directly or consumed by mankind and/or animals. They may be used as raw materials which are further processed. Plant breeders, therefore, have to follow (anticipate) all kinds of new developments in the cultivation of crops since their new varieties must be appropriate for the techniques and equipment that the grower uses. Likewise they must respond to the needs of the consumer. Sometimes these needs conflict. For example, it might be desirable to produce a thick skinned product because this would aid transport and storage. On the other hand the consumer or food processor may regard such a feature as detrimental to their needs. The point we are trying to make is that plant breeders do not only need to understand plant genetics and physiology, but they also need to have a knowledge of the objectives and practices of the grower and user.

In view of the comments we made in Chapter 2, the plant breeder must also be aware of the environmental and legal issues connected with the release of genetically modified plants. It perhaps goes without saying that the breeder also needs to be something of an economist and a business manager. It is fruitless (no pun intended!) to invest large sums of money in developing a new plant variety for which there is only a small market. Plant breeding is inevitably a complex issue. It is further complicated by the time it takes to develop a new variety. It can take from 8-15 years. The breeder needs therefore, considerable foresight to predict market demand.

3.3 The objectives of the plant breeder revisited

In Chapter 1 we briefly described the objectives of the plant breeder. They are so fundamental to any plant breeding programme it is worth re-examining these in a little more depth. Without sound objectives, considerable waste of resources can be incurred and the questions that must always be asked are:

- are the objectives for the breeding programme going to produce strain(s) of plants which will be an advance on currently available strains?

- will competitors produce as good or better varieties?

- will the demand for the product be sufficient in 10 or so years time to warrant the investment?

∏   What are the objectives of plant breeders? Perhaps before reading on, you would write down a list in the margin to see how much of Chapter 1 you can recall.

Our requirements from plants are highly variable. They range from using whole plants or parts of them to provide nourishment, to their use as raw materials for cloths, paper, spirits, tobacco etc. We also require that they are of lowest cost per unit of harvested product. This means that their cultivation costs need to be kept low. The diversity of these requirements means that we can set a large number of objectives.

polygenic
characters

Yield - This is a very important criterion. Dependent upon the crop, it may concern seed tubers, leaves, fruits, starch, sugar, oil or protein. In most cases, the yields of these products reflect the activities of a large number of genes - that is they are complex polygenic characters. For example, in wheat, grain yield depends upon the number of haulms per plant, the number of grains in each haulm and the average grain weight. Each aspect of this yield is controlled by several genes. Thus the number of haulms per plant is controlled by several genes. Similarly the number of grains in each haulm is also influenced by the activity of a variety of genes. We can anticipate therefore that to greatly influence yield, we may well have to recombine a very large number of genes. We might also anticipate that improvements in yield are more likely to be achieved by multiple gene re-arrangements (eg by conventional breeding) rather than by single gene transfer (eg by genetic engineering). Although generally true, it is not universally so. If we were able to identify circumstances in which a single gene was yield determining, then yield may be greatly influenced by single gene transfers.

polygenic and monogenic traits

Quality - This concerns characters such as flower colour, flower shape, uniformity, keeping quality, content of toxins, composition of protein, oils and starch. These characters include examples of both polygenic and monogenic traits. Recombinant DNA technology offer great potential for manipulating quality especially in such areas as flower colour, composition of storage products and keeping quality.

Earliness - In some horticultural and agricultural crops, earliness is of importance in order to capture much higher prices for the plant product. In some cases, earliness can be helpful in escaping disease.

abiotic and biotic factors

Crop Protection - Exposure to stress conditions may greatly influence the yield of a crop. Producing plants which can withstand stress is a major objective of plant breeders. We can divide stress conditions into two groups: abiotic factors like low temperatures and drought and biotic factors like diseases, pests and herbicides.

monogenic

Breeding for pest or disease resistance is very much a focus of interest of the recombinant DNA technology since in many instances resistance is monogenic, ie based on single gene products. Hardiness against environmental (abiotic) stress appears much more complex and polygenic in character.

polygerm and monogerm seeds

Cultivation costs - Mechanisation has helped to decrease cultivation costs in agriculture and horticulture. In the breeding programme of many crops, this aspect has been taken into account. For example, varieties adapted to mechanised sowing. This can be illustrated by the development of new strains of sugar beet. Traditionally sugar beet seeds are polygerm seeds, that is they are produced in small clusters. Thus sowing these types of seed meant that small clusters of seedlings were produced. These would have to be subsequently thinned out (usually by hoeing) to produce single, well spaced plants. The development of drills capable of delivering single seeds, together with the production of monogerm seeds by plant breeders led to considerable savings in the cultivation of sugar beet.

It is also in the breeder's interest to consider changes in harvesting techniques. For example, traditional varieties of beans, peas and sprouts do not all mature together. Such varieties demand that they are hand picked since the picker needs to select the ripe product from the immature. These crops need to be 'picked over' several times. Clearly the harvesting costs of such crops are high. Development of strains in which all of the product matures at the same time facilitate the development and use of mechanical harvesting devices leading to great reduction in costs.

Before we move on to examine how conventional breeding may be used to achieve some of these objectives, we will need to cover some background material relating to the language of genetics.

3.4 Some important terms in plant breeding

We will divide our description of terms into two parts:

- terms which directly refer to the genetic status of the plant;

• terms which refer to the characteristics (phenotype) of the plant.

3.4.1 Genetic terms

nucleus

We begin with the nucleus. We know that the nucleus contains the main part of the genetic information in a cell and that this genetic information (DNA) is contained within chromosomes. The number of chromosomes present in the cell is characteristic of the organism and cell type.

chromosomes

haploid

If the nucleus contains one single genome, consisting of a species specific set of chromosomes which as a group form a functional unit, it is said to be in the haploid condition. (In Man, sperm and ova are in the haploid condition and each nucleus contains 23 chromosomes, in Drosophila the haploid number is 4). If the nucleus contains a double set of chromosomes (usually one set from each parent) they are in the diploid state. This is the most common state in the somatic cells of higher plants and animals. The triploid state means that their are three sets of chromosomes present in the cell. Polyploids contain more than two sets of one genome (autopolyploid) or sets of two or more different genomes (allopolyploids). Polyploidy and haploidy are more common in higher plants than they are in animals. Aneuploidy is the name given to the condition in which a nucleus contains extra (but not complete sets of) chromosomes.

diploid

triploid

polyploidR
aneuploid

Where do the chromosomes in the nucleus come from? In most systems, one set of chromosomes is derived from a male gamete, the other from a female gamete. We can visualise for example two haploid nuclei fusing to form a diploid or two diploid nuclei fusing to form a tetraploid (How could a triploid be produced?).

homozygous

The products of such fusions are called zygotes. When the two sets of chromosomes in the zygote are identical the product is said to be homozygous. This is typical of self-fertilising systems such as barley and wheat. If the genes of the two sets of chromosomes are different (eg they come from different origins), then the zygote is said to be heterozygous. This is typical of cross-fertilisers.

heterozygous

Let us check that you have understood these terms by doing the following SAQ.

SAQ 3.1

Below is a haploid set of chromosomes.

1)   2)   3)   4)

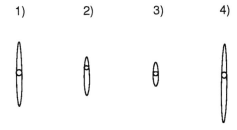

We have produced some new nuclei by fusing gametes together. The products of these fusions are drawn on the next page. Examine each and describe their ploidy status And whether or not they are homozygous or heterozygous.

SAQ 3.1

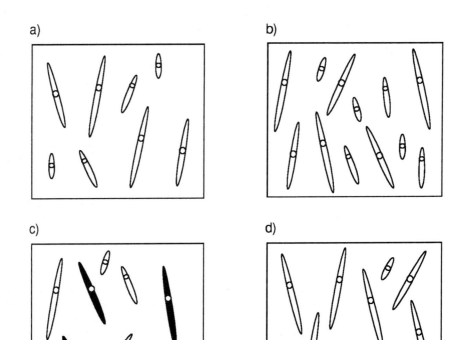

More usually the terms homozygous and heterozygous are applied to a single gene or a limited set of genes. If we take a particular gene, for example the gene that codes for an enzyme. Within the population there may be several different versions (alleles) of this enzyme. In other words there must be several different versions of the gene coding for the enzyme. We will call these different versions A1, A2, A3 etc.

Π   Now examine the diploid cells represented below:

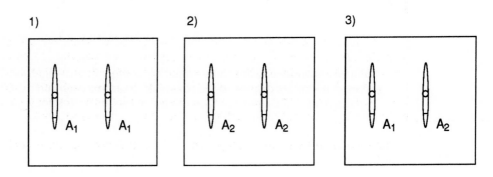

alleles

Can you see that we can describe Cell 1 as homozygous with respect to gene A? Cell 2 is also homozygous. On the other hand Cell 3 is heterozygous with respect to gene A. The different forms of gene A are said to be alleles.

dominant
recessive

Usually in heterozygous cells only one of the alleles is expressed. Thus if we return to Cell 3 drawn above we would either find the product of gene A1 or A2 but not both. In other words one would be dominant the other recessive. Let us say that A1 is dominant, we could represent this using the nomenclature Aa where A = A1 and a = A2.

3.4.2 Cultivars and genetic constitution

We have been using a variety of terms to describe the product of the plant breeder; terms such as strain, variety and cultivar. It is important to understand the significance of these terms both from a botanical stand-point and from a commercial one. To do this we begin by a brief consideration of a natural population of plants or animals. We know species we can divide them up into groups which we call species. We define species as a group of similar organisms capable of inter-breeding. Although this definition has its limitations, it is widely accepted as a convenient concept.

races

A species is not a collection of identical organisms but a collection of similar organisms. Within a species therefore we may find sub-groups (races, varieties) which show even greater similarity to each other. Thus the species *Homosapiens* (mankind) contains many varieties millions of individuals, all different (except for identical twins) from each other but which can be grouped together into races (varieties) such as Caucasoid, Negroid etc. A clone similar situation applies to populations of plants. We can recognise species, varieties (races) and individuals. A sub-population of a species consisting of genetically identical line individuals is described as a 'clone' (vegetatively propagated) or a 'line' (self-fertilisers).

cultivar

In plant breeding, the term cultivar is important. A cultivar is a group of plants which has defined characteristics. The characteristics which are used differ for different cultivated plants. Thus for example we might define a cultivar of rose in terms of plant vigour and shape, hardiness and flower colour. With other plants we may use other criteria.

∏ Is a clone a cultivar?

It follows that if all the plants in a population are genetically identical, then the plants will have the same properties. We could define such a population as being a cultivar.

∏ Are all cultivars clones?

The answer is certainly no. The genetic constitution of a cultivar is dependent on the reproduction system of the crop. In vegetatively propagated crops such as potato, *Alstroemeria* and cassava, a cultivar is homogenous and consists of one heterozygous genotype. This needs some explanation. The original cultivars were produced by cross-pollination giving rise to heterozygotes. In this case the cultivar is also a clone since although heterozygous, all the ramets are genetically identical.

Homogenous cultivars are also produced by self-fertilising crops such as barley and wheat.

Π Rye (*Lolium perenne*) propagates by cross-fertilisation. Would you expect cultivars of this crop to be homogeneous or heterogeneous? Would they be homozygous or heterozygous?

In this case the cultivars are both heterogeneous and heterozygous. Because these plants are incapable of self-fertilisation, the zygote is produced by the fusion of two different haploid nuclei. Genetically, we would expect a variety of different nuclear fusions to take place between the gametes from different individuals. The population will therefore be heterogeneous. It is still acceptable to describe such a population as a cultivar providing all of the plants fulfil the criteria used to describe the cultivar. Of course it is easier to guarantee the properties of a cultivar which consists of a homogenous population. This could be produced in the following way:

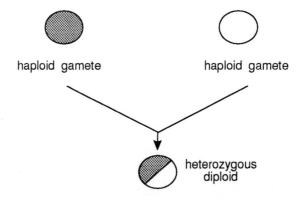

These heterozygotes can then be grown up to form mature heterozygous plants

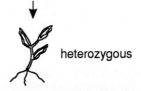

Individual plants can then be reproduced vegetatively to produce a homogenous population

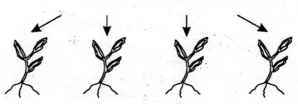

hybrid cultivars Finally we need to explain one further term - hybrid cultivars. Hybrid cultivars are produced by fusion of the gametes of two definable genotypes. Consider the simple case of a plant whose haploid number of chromosomes is two and let us assume that each gene exists in two allelic forms. Thus eight possible diploid genomes exists. Namely:

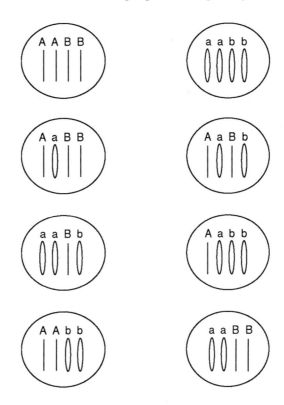

We could represent them as AABB, aabb, AaBB, AaBb, aaBb, Aabb, AAbb, aaBB.

In hybrid cultivars, two homozygous strains are crossed to give predictable progeny. Thus:

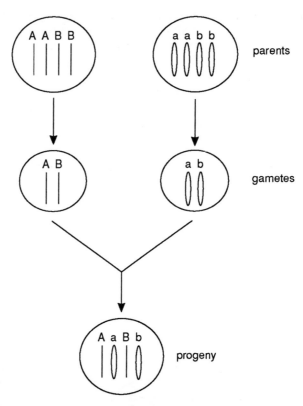

We could call the progeny a hybrid variety.

Π   If we started with parents AaBB and AABb, would all the progeny have the same
     genotypes? (Draw out a flow diagram like that described above to help you solve
     this).

You should have come to the conclusion that the answer is no. The first parent (AaBB)
can produce two types of gametes (AB and aB). The second parent would also produce
two types (AB and Ab). Thus the offspring of such a cross would contain a mixture of
genotypes (ie AABB; AaBB; AABb; AaBb) depending on how the gametes fused.

Π   Let us repeat this with the parents of genomic composition AAbb and aaBB.
     Would all of the progeny have the same genetic composition?

The answer is yes because both parents can only produce one type of gamete (ie Ab and
aB respectively). Therefore there is only a single type of offspring (hybrid) possible -
AaBb. If this hybrid was allowed to interbreed would the subsequent offspring be
genetically all the same? (Think of the gametes that could be produced). You should
come to the conclusion that the next generation will contain individuals with the
genotypes of: AABB; AaBB; AaBb; Aabb; aaBB; AAbb; Aabb; aabb.

Hybrid cultivars are important commercially. Many of these hybrids show much
improved performance (yield, quality etc.). The fact that interbreeding of the hybrid will
quickly generate gene combinations that are not so desirable means that the hybrid

cultivar has to be continually generated from appropriate parents. Of course this places the plant breeder who has the appropriate parents in a strong position compared to the plant grower who purchases the hybrid seed.

SAQ 3.2

A single diploid plant, grown from the seed produced from a cultivated plant which had been cross pollinated by a related wild plant showed several desirable characteristics. This plant, however, produced infertile flowers and had to be propagated by taking cuttings.

Which of the following phrases can be properly applied to the population of plants produced by this type of vegetative propagation?

1) The population would be genetically homogenous.

2) The population would be homozygous.

3) The population could be regarded as a clone.

4) The population should be regarded as a new species.

5) The population could be regarded as a cultivar.

Now that we are understood some of the language of plant breeding, let us turn our attention to the strategy that is adopted to produce new cultivars of valued crops by conventional breeding programmes.

3.5 The strategy of the conventional plant breeder

Because of a large variation in breeding factors, crops, cultivation methods and ways of propagation, it is impossible to develop a universal breeding approach. However, in the breeding process a few general steps are recognised:

breeding objectives

Step 1. The first stage is to establish the objectives of the breeding programme. These may be general or very specific. It is essential however, to establish these objectives early on since they will govern the remaining steps.

selection of starting material

Step 2. This stage focuses on the collection and selection of the plant material that acts as the starting point for the experimental part of the programme. It usually involves selecting plants which display one or more of the desired characters. For example if the objective was to produce a high yielding wheat which was resistant to certain diseases, the usual starting point would be to select a collection of high yielding wheat and a collection of related plants which were disease resistant.

crossing

Step 3. Crosses are made between the selected starting material in an effort to combine the desired characters by recombination. This is one of the more restrictive phases of conventional breeding programmes because successful recombination producing fertile off-spring only occurs with quite closely related plants. Therefore it is only possible by these techniques to recombine genes from closely related sources.

selection

Step 4. Within the variation obtained in the offspring of such crosses, the desired new phenotypes have to be selected and one of the promising individuals, lines or populations have to be developed into the new cultivar (ie large numbers produced). The way this is achieved depends mainly on the method of propagation (vegetative, sexual, mankind-dependent, mankind-independent) and of the properties sought in the new cultivar. Before the cultivar is sold it needs to be field tested.

disease-free
prevention of
inbreeding

A cultivar then has to be maintained, propagated, tested and sold. The maintenance of cultivars is therefore a very important activity of the breeder. The problems involved are strongly connected with the method of propagation. In vegetatively propagated crops, like potato, disease-free maintenance of 'seed' potatoes tubers is essential. In hybrid cultivars contamination with undesirable inbred seed has to be prevented (see previous section).

∏ It would be useful to produce a flow diagram to remind yourself of the stages described above. Use the activities listed in the box below to produce a flow diagram of the sequence of events which lead to the marketing of a new cultivar. (We have mixed them up in order to get you to think carefully of the order in which they should be carried out - use the description given above to help you).

> Selection of progeny; selection of parents; set objectives; propagate progeny; cross-fertilise; market cultivar; field test.

3.6 Variation heritability

In Section 3.5 we mentioned that there is variation in the offspring derived from crosses. This variation may be genetic or phenotypic or both.

Consider the situation of two homozygous diploid plants, one produces red flowers, the other white flowers and that flower colour is controlled by a single allele. Also let us say that the red allele is dominant. We could represent the homozygous plant producing red flowers as having the genetic composition of RR where R represents the dominant red pigment-producing gene.

locus

The homozygous plant producing white flowers would have the recessive gene (r) at the same position (locus) on the equivalent chromosome. So we could describe this as having the genetic composition of rr.

Diagrammatically:

red flowered parent   white flowered parent

If we produce gametes from these, they would have the following composition:

The diploid progeny, (the so called F_1 generation) produced from fusion of these gametes would have the composition of:

F_1 progeny

They would all produce red flowers.

If we allowed these to interbreed, they can form the following gametes:

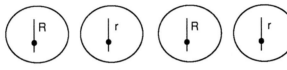

which could fuse in the following combinations:

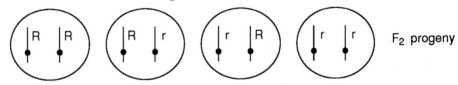

F_2 progeny

The first three products of this generation (F_2 generation) would produce red flowers, the last one would produce white flowers.

This is, of course, the classic case of Mendelian inheritance. The point we are trying to make is that we get phenotypic variation in the F_2 generation. This type of inheritance pattern where the variability is discontinuous (ie white or red but no intermediate shades) is typical of qualitatively inherited variability. It is most clear cut where a single gene qualitatively influences a phenotypic character.

qualitatively inherited variability

quantitatively inherited variables

In quantitatively inherited characters such as yield there is a continuous variation between individuals. Typically these characters are polygenic thus offering a large number of possible combinations of the genes contributing to the character. Further variability is introduced by the influence of environmental factors on the population of plants.

measurement of variability

It is useful for plant breeders to be able to measure variability and to distinguish between the variability encoded within the genes of a cultivar and the variability produced by the influence of environment.

The phenotypic variance we can measure. This is often given the symbol Vp. We can split this into two main components - genetic variance (Vg) and environmental variability (Ve). Thus if we can write:

Vp = Vg + Ve.

heritability   Heritability is given the symbol H and specifies the proportion of the total variability that is due to genetic causes. Mathematically we can write the relationship:

$$H = \frac{Vg}{Vp} = \frac{Vg}{(Vg + Ve)}$$

If we grew the plants in a carefully controlled, uniform environment, Ve would of course be close to 0. In this case all of the variability observed in the population would be due to genetic causes. In this case H would = 1. If on the other hand we grew a population of plants which represent a single clone, Vg would be 0 and all of the variation in the population would be a consequence of the environmental factors. In this case H = 0.

SAQ 3.3

Which of the following phenotypic characters would be likely to show quantitatively inherited variation?

1) Resistance to a particular viral infection.

2) Cold tolerance.

3) Plant size and shape.

4) Resistance to a single herbicide.

5) The number of seeds and the size of seeds produced in the fruit.

SAQ 3.4

A clone of lettuces are grown in a thermostated evenly illuminated growth cabinet with air circulators.

1) What is the anticipated heritability (H) of the variation observed in such plants?

2) What is the anticipated phenotypic variation (Vp) for such plants?

3) If the plants were grown in a garden surrounded by a wall on one side and a path on the other, what would happen to the Vp value?

4) Given the condition described in 3), what would happen to the H value?

Now that you have completed these questions you can probably understand why it is important for a breeder to have some measure of the heritability of the variation. If the heritability (H) of variability is 0 then it is of a waste of time to use a procedure to select one of the variants showing the most of a desired character to act as parent for genetic crossing experiments. The desired character is not inherited but a consequence of the environment in which the plant was grown.

In other words if we had a plot of land on which we were growing carrots, selecting the largest may not be a good strategy for choosing parents for genetic crossing. The size of the carrot may be a product of its position in the plot rather than of any inheritable traits. Thus knowing the value of H may influence the strategy of selecting parents. Of course it makes sense to minimise Ve by growing the plants under very similar (identical) conditions so that any observed variability is likely to be due to genetic variability.

For a similar reason, variability in the progeny of a cross is important. Again the breeder seeks to identify the variants which have inherited features and not to use the variants which arise simple through environmental factors.

conditions favouring high and low H values

From the point of view of the grower, the heritability of variation is also important particularly in, for example, trying to produce even sized plants or flowers, or to have plants ready for use at a particular time or over a period of time. Here the small gardener and the large producer may be looking for quite different conditions. Home growers do not want all of their cauliflowers ready for cutting within a few days of each other - many would go to waste. On the other hand, there are many advantages to commercial growers having all of their crop ready at the same time - it would reduce harvesting costs if each field had to be visited only once for harvesting.

3.6.1 The search for genetic variation

quantitative variants more difficult to find than qualitative variants

We have identified two types of variation, qualitatively and quantitatively inherited variation. It should be obvious, because of its all or nothing effect, that qualitatively inherited traits are the easiest ones to deal with. The search for quantitatively inherited variations need much more attention. Firstly, it is important to recognise whether or not the variation is a consequence of environmental or genetic factors. Secondly, these variations are not all or nothing effects, but are smaller quantitative differences. Thus careful quantification is essential.

∏   In looking for desirable genetic variants, where is the plant breeder to look? Below we have provided a list. See if you can write a reason why we have listed each in the order we have.

Sources of plants exhibiting genetic variability:

* existing modern varieties;

* spontaneous and induced mutants;

* primitive races/uncultivated material from centres of origin;

* related, mainly uncultivated species preserved and maintained in gene banks.

We placed existing modern varieties at the top of the list not because they are likely to show more variation than the others, in fact they probably show less because of inbreeding. But such strains have already been developed to produce many of the desired characteristics. Going back to, for example, uncultivated material from centres of origin may well bring in wider genetic variation and new genes, but these uncultivated plants will not have been developed for cultivation. In other words they would have been selected for by evolutionary rather than agricultural pressures.

spontaneous and induced mutations

importance of gene banks

We have ranked mutation fairly high because mutations are a direct route to genetic variability. It occurs naturally (spontaneously) or can be induced in a variety of ways. We will examine this source of genetic variability in a later section. We ranked preserved materials in gene banks quite lowly because this source suffers from the problem of needing to be resuscitated and the draw-backs highlighted in the above discussion on uncultivated sources are also prevalent. If, however, a plant with a particular, desired trait is known to exist in a gene bank then this becomes a valuable source. Gene banks of course offer a way of insuring potentially valuable traits (genes) are not lost.

3.6.2 The scope of genetic recombination by conventional breeding

restricted to
closely related
plants

Conventional breeding normally takes place within groups of closely related plants. Usually the plants used are members of the same species. Conventional breeding programmes are limited by the ability of the pollen of one plant being able to fertilise another and by the ability to grow the recombinant progeny. Thus conventional breeding programmes are largely confined to the genetic variation found within a group of plants capable of interbreeding. There are however techniques which may be applied to expand the genetic variation available to breeders. Perhaps the most impressive of these is the introduction of genetic engineering techniques which, in principle, enable genes from a wide variety of sources to be transferred into a plant. These modern biotechnological approaches even enable the development of synthetic genes. The ways of broadening the genetic variation available to the breeder are by using:

- induced mutation;

- inter-specific hybridisation;

- genetic engineering (recombinant DNA technology);

- somatic hybrisation.

We will examine these in more detail in later sections. The main message we wish to convey is that essentially what plant breeders do is to recombine the genetic variation available to them and, to produce combinations which exhibit desired traits.

expanding the
pool of
available genes

Thus if we think of the pool of genes that is available in a population of interbreeding plants - the pool can be expanded by both traditional techniques (eg induced mutations) and the more contemporary procedures (recombinant DNA technology and somatic hybrisation). These latter, biotechnological procedures, may therefore be integral parts of more conventional breeding programmes.

3.7 Techniques for expanding genetic variation available to the plant breeder

3.7.1 Induced mutation

replacement

deletion

addition

Mutations are based on changes in the nucleotides of DNA. They may be the product of replacing one type of nucleotide with another; they may be the product of removal (deletion) of nucleotides or the addition of extra nucleotides. Mutations can operate at a single nucleotide level (ie at an individual gene level) or may involve parts of chromosome, whole chromosomes (aneuploidy) or even whole genomes (polyploidy). The feature of mutations is that by changing the nucleotides, the information stored in DNA is also changed. Depending on the nature of these changes this may give rise to changes in the phenotypic traits of the plant.

Thus the process of using induced mutants to expand genetic variation can be described by the sequence drawn in Figure 3.1.

Π In producing mutants by the process described in Figure 3.1, what are the key issues as far as the plant breeder is concerned?

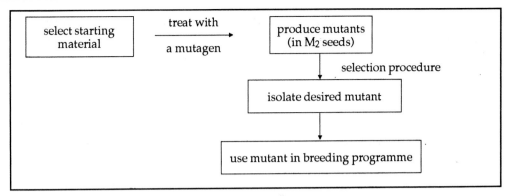

Figure 3.1 The production of mutants for use in plant breeding programmes.

selection of mutagens

isolation of mutants

The types of mutants that are produced depends very much on the nature of the mutagen (agent which induces mutants) used. These mutagens include a variety of chemicals such as the alkylating agents and physical agents (eg UV light) and other high energy radiations. If you are unfamiliar with these agents we would refer you to the BIOTOL text on "Genetic Organisation in Prokaryotes" or to any good basic genetics text. In addition to selection of the mutagen, the other key issue is to develop a suitable screening procedure in order to select and isolate the desired mutants from the (M_2) progeny.

empirical

This type of approach to increasing the genetic variation available to the breeder is largely empirical. Mutagenic agents usually act randomly on the genome producing effects in different genes in different plants (M_1 - seedlings). Thus the plant breeder takes a large amount of starting material, treats it with a mutagen and then has to screen a large number of plants to identify those with the desired traits. Once such a mutant has been isolated it can, by cross breeding programmes, be transferred into a variety of cultivars.

extra-chromosomal inheritance

Before we leave the issue of induced mutants, it is important to recall that not all of the plant genome is found in the nucleus. DNA is found in chloroplasts (cpDNA) and mitochondria (mtDNA). Induction of mutants in these extra-chromosomal DNAs offer some advantage to the plant breeder. They are expressed independently of the ploidy level of the plant and they are inherited maternally. Examples of extra-chromosomally inherited mutations are herbicide (t) resistance encoded in the cpDNA of *Brassica oleracea* and cytoplasmic male sterility in the mtDNA of onion.

The chromosomally encoded mutations may be dominant or recessive. Let us check that you understand what this means by doing SAQ 3.5.

Π How can recessive mutants be identified? (Use SAQ 3.5 to help you).

Clearly to get expression of a recessive gene, we have to produce a homozygote for the relevant allele (see example 2 in SAQ 3.5). For diploid plants, this is most easily achieved with self-fertilisers (pea, tomato). With normal cross-fertilisers such as maize we have to deliberately self-fertilise the plant.

SAQ 3.5

In the diagrams below, we have labelled the mutant allele as Nm and the normal gene as N. For each of the diploids we have illustrated, write down the gene product that will be produced (ie either Nm or N).

	Diploids		
1) Nm is recessive	N Nm	Gene product	[ ]
2) Nm is recessive	Nm Nm	Gene product	[ ]
3) Nm is dominant	N Nm	Gene product	[ ]
4) Nm is dominant	Nm Nm	Gene product	[ ]

Π   Draw a diagram to explain how homozygotes can be isolated from heterozygotes by self-fertilisation (If you have difficulty, look back to Section 3.6 and examine the diagram which describes the crossing of F_1 generation hybrids to form F_2 progeny. Examine the combination of genes in the progeny).

In plants capable of self-fertilisation many different mutants have been isolated and investigated, but only a relatively small number have found practical use.

Examples of such mutants are semi-leafless mutants of pea and amylose-free starch in waxy mutants of maize and rice.

It is almost impossible to obtain homozygous recessive segregants after mutant induction in cross-fertilising and polyploidal crops. This is a real handicap for mutation breeding in these crops. One way is to use haploid cell cultures. Haploid cultures can be obtained by culturing anthers. The production of haploid cell lines from anthers is described in the BIOTOL text 'In vitro Cultivation of Plant Cells' (see also Section 4.3.5).

3.7.2 Interspecific hybridisation

Resistance against diseases is found in related species of many crops. The way of transferring such genes into the cultivated plant is dependent on the degree of relationship. When there is a choice out of different sources the preferred order is: modern varieties - land races - uncultivated material - related species - distantly related species. When distantly related species are used, pairing between homologous chromosomes can be disturbed, preventing introgressive recombination (NB Introgression is the name given to the introduction of genetic material from one gene pool to another by recombination in interspecific hybrids). Gene transfer between different species can be obtained by three main methods:

introgression

•   normal introgressive recombination;

•   induced introgressive recombination by translocation;

•   substitution of cytoplasm.

Normal introgressive recombination

Hybridisation between the crop plant and a related species (assuming there are no major problems of infertility) is followed by re-crossing with the parental crop species so that the resistance gene of the related species becomes transferred to the crop species and that undesired genes of the related species are eliminated.

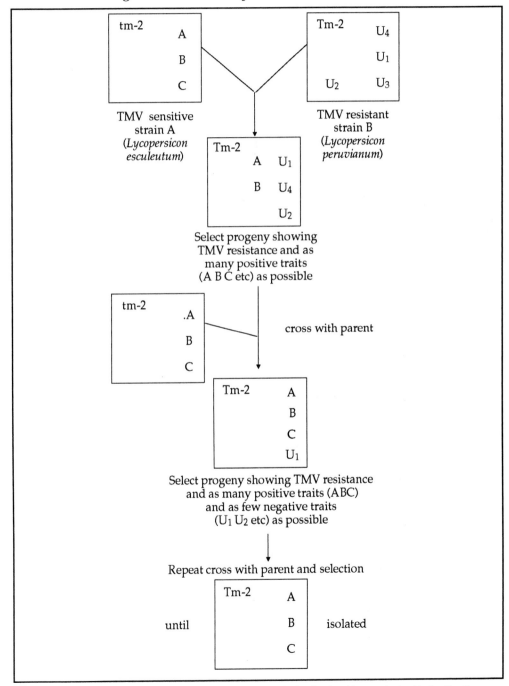

Figure 3.2 Stylised introgressive recombination. A, B, C = desired traits , U_1 = undesired traits, Tm-2 = TMV resistance gene.

In Figure 3.2 we use the transfer of the Tm-2 gene (resistance to tobacco mosaic virus) from *Lycopersicon peruvianum* into tomato (*L. esculeutum*) as an example. The main problem is to remove undesired characters (labelled U1 U2 etc) which are simultaneously transferred with the Tm-2 gene. This is done by back-crossing with the parent cultivar. The scheme shown in Figure 3.2 is somewhat simplified and it may take many back cross generations before the desired strain is achieved.

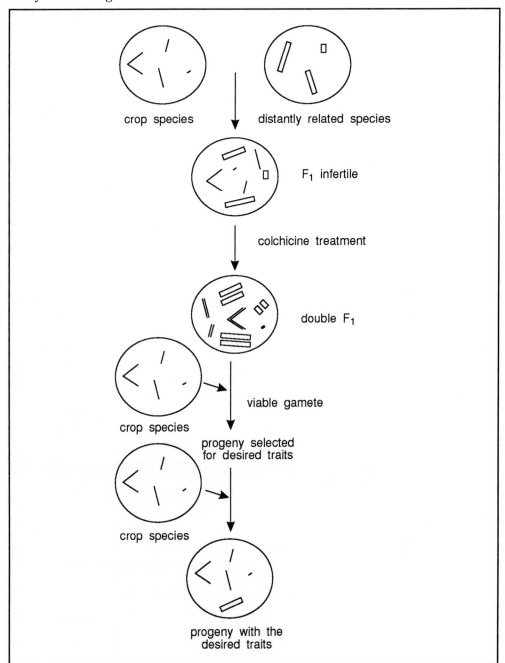

Figure 3.3 Induced introgressive recombination (see text).

Induced introgressive recombination by translocation

A resistance may be found in a distantly related species which is still crossable with the cultivated plant. However, if the homologous chromosomes do not pair during meiosis the progeny F_1 is sterile because gametes do not contain all of the essential chromosomes.

colchicine treatment

In a few cases this can be overcome by deliberately interfering with meiosis by using colchicine. Colchicine prevents mitotic spindle formation. Thus if colchicine is applied, a double plant can be obtained. This means that the gametes that are produced contain all of the genetic information that is required for viability.

Re-crossing (back-crossing) of these F_1 plants with the parental crop species under continuous selection for the desired character enables isolation of progeny with a complete set of crop plant chromosomes plus one extra chromosome from the wild species carrying the desired character. We have represented this process in Figure 3.3.

Note that the product contains a complete set of crop plant chromosomes and one extra chromosome from the distant relative. By treating this product with high energy radiation, it is possible to translocate part of the chromosome. Thus:

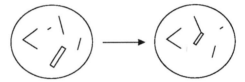

By careful selection of progeny derived from the irradiated material, a crop plant carrying a few (including the desired) genes from the distant relative may be isolated.

Some successes have been obtained (Table 3.1) but this method is:

- laborious;
- only applicable to simply inherited characters;
- the piece of chromosome with the resistance gene is randomly integrated into the genome of the recipient (meiotic abnormalities are expected to occur after crossing the translocated line with normal varieties);
- the inserted part of the chromosome generally carries undesired genes of the donor species.

⨅ Which is the most common type of high energy radiation used to facilitate translocation? (You can obtain an answer from Table 3.1).

Alien species	Pathogen resistance	Method used	Cultivars produced
Aegilops umbellulata	*Puccinia recondita*	X-rays	Riley 67 Arthur 71 etc
Agropyron elongatum	*P graminis*	Gamma-rays	Eagle Kite
Secale Cereale	*P graminis*	X-rays	-
Secale cereale	*P.recondita*	X-rays, neutrons	-
Agropyron elongatum	*P.graminis,* *P.recondita*	X-rays, neutrons	-
Agropyron elongatum	*P.graminis,* *P.recondita*	Spontaneous translocation	Agent etc
Secale cereale	*P.striiformis,* *P.recondita,* *P.graminis*	Spontaneous translocation	Aurora,, Kavkaz, etc
Agropyron intermedium	All 3 rusts	Radiation	-
Aegilops speltoides	*P.recondita*	Cross	-

Table 3.1 Genes for rust resistance that have been transferred to wheat from distantly related species.

Substitution of cytoplasm

maternally inherited cytoplasmic inheritance

cytoplasmic male sterility

Substitution of the cytoplasm of the cultivated species by that of another species can be obtained through recurrent crossing of the F_1 plants. After using the species as female parent, no cpDNA and mtDNA of the cultivated male parent is transferred. For breeding of hybrid varieties; exchange of cytoplasm is of great importance when it is coupled with cytoplasmic male sterility (cms). This is the result of an imbalance between nucleus and cytoplasm. Cms induced through substitution of cytoplasm is possible in several crops like wheat, cotton, sunflower, rapeseed. Cms occurs spontaneously in onion and maize. In *Brassica napus* cms is induced by the cytoplasm of *Raphanus sativum*. However, this source of cms is connected with sensitivity to cold stress encoded by cpDNA of *R. sativum*. Therefore, its application is restricted.

SAQ 3.6	Explain why producing cytoplasmic male sterility is important in breeding hybrid varieties.

3.8 Mode of reproduction and its genetic consequences

For breeding of new varieties the selection process is a key activity. This is closely connected with the reproduction of the plant. Cultivated crops can be classified as:

- vegetatively propagated crops (potato, cassava, *Alstroemeria*);

- self-fertilising crops (pea, *Phaseolus*, beans, wheat, tomato);

- cross-fertilising crops (rye, maize, sugar beet);

- directed cross-pollination enabling the production of hybrid varieties.

The genetic constitution of varieties obtained, is dependent on the method of propagation as shown for diploid crops in Table 3.2.

Crop	Plant	Propagation	Genetic constitution
Barley	1	self-pollination	AAbbccDDEE
	2		AAbbccDDEE
	3		AAbbccDDEE
Rye	1	cross-pollination	AABbCcDdee
	2		AaBbccDDEe
	3		AaBBCcDdEe
Apple	1	vegetative	AaBbCcDdEE
	2		AaBbCcDdEE
	3		AaBbCcDdEE
Hybrid maize	1	crossing 2 lines	AaBbCcDdEE
	2		AaBbCcDdEE
	3		AaBbCcDdEE

Table 3.2 Genetic constitution (stylised) of modern varieties in several diploid crops produced by a variety of propagation procedures.

∏ Examine Table 3.2 carefully. Which types of propagation favours the production of homozygotes? Which types of propagation promote heterogenity in the progeny? Which types of propagation promote homogenuity in the progeny?

You should have come to the conclusion that self-pollination favours the production of homozygotes. Heterogenity is favoured by cross-pollination. Homogenity of progeny is favoured by self-pollination and by vegetative propagation. It is also a feature of the F_1 generation of hybrid crops.

3.8.1 Vegetatively propagated crops

After vegetative propagation using plant parts such as bulbs (tulips), tubers (potato) or rhizomes (*Alstroemeria*) the offspring are genotypically identical and constitute clones. At the end of the selection process a superior clone is the basis of a variety. The majority of vegetatively propagated crops are strongly heterozygous. The greatest problems in vegetatively propagated crops are producing disease (virus) free plants, maintenance and propagation.

Here modern biotechnological procedures have greatly aided the plant breeder. The introduction of plant and cell tissue culture methods and micropropagation procedures enable the maintenance and production of large numbers of plants in disease free environments.

3.8.2 Self-fertilisation

Self-fertilisation leads to homozygosity. This is shown for one locus in Figure 3.4. Note that if we begin with 100% heterozygotes (ie an F_1 generation), after one self-fertilisation, the proportion of heterozygotes falls to 50%. By the F_4 generation it will fall to about 12.5%.

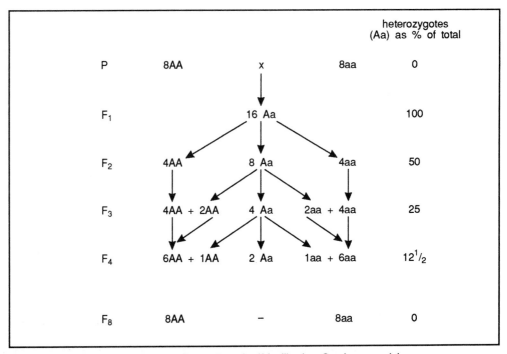

Figure 3.4 Degree of heterozygosity after continued self-fertilisation. One locus model.

⊓ What would the proportion of heterozygotes in the F_8 generation? (Have a go at this calculation - you should find out that the answer is <1%).

3.8.3 Cross-fertilisation

promotion of
heterozygosity

Cross-pollination is common in higher plants. All kinds of systems are developed promoting cross-pollination and preventing self-fertilisation such as in dioecious (plants in which each plant produces male or female flowers but not both) and self-incompatible plants. By cross-pollination, heterozygosity is promoted and preserved. The progeny of such a plant is genotypically a population containing a great number of genes with different alleles which are exchanged. It is relatively simple to select for true breeding characters which are monogenic recessive. Selection of a dominant trait is more complicated because of the presence of AA and Aa genotypes in one phenotypic class. Mass selection and the procedures of progeny selection and line selection are more commonly used than single plant selection.

3.8.4 Hybrid varieties obtained by directed cross-pollination

heterotic growth

A hybrid variety is an F_1 population, obtained by crossing clones, inbred lines or other populations that are genetically dissimilar, that is used for commercial plantings. Such varieties make, like in vegetatively propagated varieties, better use of heterotic growth (ie show hybrid vigour). When inbred lines are used, heterozygous hybrid varieties are phenotypically homogeneous with genetically identical individuals. Of course this is only true if the parents are so inbred that they are homozygous at all locci. As we have learnt earlier, inbreeding produces segregation of many characters in the next generations. Therefore, hybrid varieties protect breeders against the use of their own harvested seed by the grower. Hybrid varieties are very popular and can be made in self-fertilising or cross-fertilising crops.

3.8.5 Self-fertilising crops

cost of producing hybrids and the value of the product

In tomato, hand-produced F_1-hybrids are very popular because of uniformity, heterotic growth and breeder's protection. Per pollination a few hundred seeds are obtained. This is economical because the grower is willing to pay a higher price for hybrid seed. However, in wheat and barley hand-pollination of hybrid seeds production is uneconomical. In such crops a reliable genetic (male sterility) or chemical (gametocides) emasculation method is needed for seed production.

3.8.6 Cross-pollinators

use of cytoplasms

male sterility to prevent in-breeding

In maize, sugar beet and *Brassica* species, hybrid varieties are popular because of hybrid growth and homogeneity. The production of fertile in-breeding lines is a problem. In maize and sugar beet cytoplasmic male sterility is used. Thus F_1 seed production is prevented from in-breeding. In *Brassica* species grown in horticulture (eg cauliflower and Brussel sprouts) incompatibility is used. However, expression of incompatibility is often environment dependent and, therefore a problem. Inbred lines are made by hand-pollination of flower buds. Breeding programmes using this technique are resource demanding.

3.9 Plant biotechnology and conventional breeding

During the last 20 years a lot of progress has been made in the fields of cell biology and molecular biology. This has had its influence on breeding research. Breeding research at cellular and molecular level is better known by the term plant biotechnology. Plant biotechnology is the application of cell biological and molecular biological knowledge and techniques in plants with economic goals. We will deal with these techniques and their application to plant breeding in greater depth in subsequent chapters. Here we will provide an overview of how biotechnology may contribute and interact with conventional breeding programmes.

We can broadly divide biotechnological contributions to plant breeding into three main areas namely:

- cell biological aspects;

- molecular biological aspects;

- genetic manipulation.

If we think back to conventional plant breeding, we will recognise that plant breeders are faced with the three following issues. Firstly they attempt to use the maximum possible range of genetic variation that is available. Secondly they have a need to be able to select the recombinants that show the desired characteristics. Thirdly they have a need to generate large numbers of progeny with predictable properties (eg virus-free). How then may the biotechnological developments in cell biology, molecular biology and genetics contribute to the plant breeder's programme? We will deal with these in the order described above.

3.9.1 Cell biological aspects

in vitro
cultivation

Recent developments in the *in vitro* cultivation of plant materials (eg multiplication of cells, plant regeneration from explants, callus growths) and the fusion of protoplasts, anthers, microspores and cells are all important to the plant breeder. For example *in vitro* cultivation enables production of large amounts of genetically identical plant material. It also provides suitable material for genetic engineering. Protoplast fusion between irradiated donor protoplasts and untreated recipients is an alternative approach to partial genome transfer. In other words, this increases the diversity of genes available for the plant breeder. It also enables new combination of chromosomal and cytoplasmic DNA to be created.

cybridisation

It is generally accepted that fusion products contain chloroplast cpDNA and mitochondrial mtDNA from one of the fusion partners. Therefore, somatic (protoplast fusion) hybridisation is also a means to induce, in one step, cytoplasmic substitution of extra-chromosal DNA. This is called cybridisation. In cybridisation products, the recipient contains its own nucleus but mtDNA and/or cpDNA of the donor species.

3.9.2 Molecular biological aspects

The importance of molecular biology for plant breeding comes from the fact that all kinds of DNA sequences can be isolated and that vector plasmids and vector systems are developed allowing introduction of foreign genes into the plant. The molecularly isolated DNA sequences originate from plants, animals, bacteria and even viruses. We can transfer genes into plants which code for:

- storage protein in seed;

- enzymes like nitrate reductase, chalcon synthase or granule-bound starch synthase (GBSS) allowing amylose synthesis;

- compounds toxic for certain insects;

- coat proteins from viruses producing protection against viral infection;

- male sterility.

use of cDNA to
produce
genetic maps
and to identify
genes

cDNA probes (DNA copied from mRNA) of low copy or single copy genes show different hybridisation patterns with digested DNA of different plant genotypes. In potato, tomato and maize, many of these cDNA probes have been used to produce genetic maps. These maps can be integrated with the classical map containing morphological and agricultural traits (we will examine this aspect more fully in a later chapter). We can therefore use cloned cDNA probes to identify the presence of genes. For example, let us say we have cDNA for a gene linked to a certain disease resistance. We can use this cDNA to identify indirectly those plants in a progeny which are carrying the disease resistance gene. Only a small part of a plant needs to be used to carry out the cDNA challenge. Therefore we can speed up the selection of the desired parents and progeny if we use such cDNA probes.

3.9.3 Genetic manipulation

extension of
gene pool

We shall learn in later chapters that genetic manipulation demands plant regeneration from explants and/or protoplasts, molecular isolation of genes and the development of vector systems. It also demands that we can apply selective pressure to eliminate normal cells and promote the growth of genetically modified cells. The main contribution of genetic manipulation is that it greatly extends the gene pool available to the plant breeder.

SAQ 3.7

We have defined a hybrid variety as an F_1 population obtained by crossing clones or inbred lines. List as many ways as you can to show how biotechnology might contribute to the production of a hybrid variety.

Summary and objectives

In this chapter, we have provided you with a brief resume of the objectives of the plant breeder and thought about the constraints placed on the breeder restricted to the use of conventional breeding procedures. We have learnt that biotechnology extends the pool of genes available to the breeder, may provide tools for selecting desired products in enabling the breeder to produce cloned materials and to provide advantages in propagating plants meeting particular specifications. In the next chapter we begin our in depth examination of this application of biotechnology to plant breeding by examining the use of plant tissue culture techniques. To do this, we have made special reference to potatoes to show how their techniques have been applied to specific problems.

Now that you have completed this chapter you should be able to:

- describe the main obejctives of plant breeding;

- use a wide variety of terms used in plant breeding relating to the genetic status of plants;

- identify the ploidy state of cells;

- produce a flow diagram describing the main stages in a plant breeding programme;

- explain what is meant by genetic variability and heritability and to calculate heritability;

- describe the range of genetic variability available to the conventional plant breeder and how biotechnology has extended this range;

- explain in outline, how biotechnology may contribute to a conventional breeding programme.

Plant tissue culture

4.1 Introduction   66

4.2 The potato plant (*Solanum tuberosum* L.)   68

4.3 Overview of the possibilities and methods of plant tissue culture   70

4.4 Production of chemical compounds by cultured cells   88

4.5 Closing remarks with regard to the methods and possibilities which
     arise from *in vitro* plant culture   89

4.6 Application of *in vitro* techniques in the multiplication of the potato
     plant (*Solanum tuberosum* L.)   89

Summary and objectives   97

Plant tissue culture

4.1 Introduction

4.1.1 General introduction

definition of plant tissue culture

The term plant tissue culture is used to describe all types of aseptic plant culture procedures, pertaining to the growth of:

- plant protoplasts (which are 'naked' cells, from which the cell walls are removed);
- complete, but loose cells;
- tissues (such as apical meristems);
- organs (such as buds);
- embryos;
- plantlets.

in vitro

in vivo

Because the growth takes place in a sterile, artificial culture environment, we call these techniques *in vitro* techniques. Correspondingly, we can call the natural, non-sterile conditions *in vivo*. There are also conditions, which are in-between, for example screenhouses in the open field or glasshouses.

disease free

Each of these types of culture starts with a piece of plant. The ultimate goal is to regenerate an '*in vitro* plantlet' from the plant piece. Such an *in vitro* plantlet is an aseptic, small plant with a distinct root and shoot system, completely developed *in vitro* and entirely free of diseases.

explant

plant regeneration

micro-propagation

The piece of plant used as the start of the new individual can greatly vary in size (from one cell to an entire stem cutting, consisting of a number of leaves, their sub-tended buds and several stem internodes). This excised piece of plant is called an explant. It must go through a process of growth and differentiation in order to form a new plantlet. This process is called 'plant regeneration'. After plant regeneration the *in vitro* plantlet is available for direct use, storage, further *in vitro* multiplication known as micropropagation or breeding purposes.

Plant tissue culture is not a science. It is merely a range of techniques, in which the knowledge of many different scientific disciplines are combined and used. The techniques vary greatly in complexity. They are valuable alternatives when:

- stocks must be made free of diseases;
- propagation procedures or breeding must be speeded up;
- storage of germplasm (ie the whole of hereditary characters of a genotype) is required;
- the genotype must be manipulated in a way that is impossible when carried out *in vivo*.

Cell cultures may even prove to be useful for the production of desired plant compounds. Table 4.1 shows a survey of the current and future applications of plant tissue.

Current applications	Developing technologies
Pathogen elimination	Modification by mutation and somaclonal variation
Micropropagation or (clonal) rapid multiplication	Plant gene transfer by protoplast fusion Gene introduction by various techniques
Conservation and transport of germplasm	
Embryo rescue	
Haploid production (microspore, anther and ovule culture) and ploidy manipulation	
Somatic embryogenesis	
In vitro selection	
Production of chemicals by cultured cells	

Table 4.1 Current applications and developing technologies of plant tissue culture.

∏ Circle those applications listed in Table 4.1 which enable a plant breeder to use a wider pool of genetic variation. (You will have a chance to check your answer later).

In this chapter we examine the various applications of tissue culture techniques. At the end of the chapter we will examine the application of these techniques to the potato so that we can provide an in-depth treatment. Many of the strategies we describe are directly transferrable to other crops.

4.1.2 Use of plant tissue culture

The use of plant tissue culture is rapidly expanding. In The Netherlands alone there are 67 companies producing 62 million micropropagated plants per year (data of 1988). Similar micropropagation companies exist in many countries of the world. Most important countries are Belgium, France, Germany, Japan, Israel, Italy, The Netherlands, Spain, the UK and the US. For some food crops (eg the potato), plant tissue culture is very popular, even in developing countries. If you are interested in statistical data relating to commercial micropropagation in Western Europe, a report by R L M Pierik (1990 - in P C Debergh and R H Zimmerman 'Micropropagation of horticultural crops' published by Kluwer Academic Publishers, Dordrecht, The Netherlands) would prove to be useful.

high cost of planting material

effects on yield

The reasons for this expansion are obvious. Use of planting material of high quality is very important. In developing areas, the most costly input in food production is often planting material, especially when it must be bought on the international market. Poor quality seed can be a major yield constraint. A good example is provided by the potato. Figure 4.1 shows that there is a close relationship between the portion of the area planted with certified seed (ie seed with a guaranteed quality) and the yield achieved.

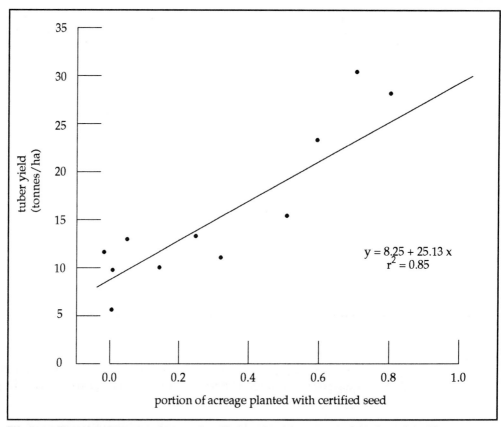

$$y = 8.25 + 25.13\,x$$
$$r^2 = 0.85$$

Figure 4.1 The relation between the use of certified seed and the average yield of potato in different regions of the world derived from Themawerkgroep Aardappelen, 1986 - Ontwikkelingssamenwerking en aardappelen, Ministerie van Landbouw en Visserij, Den Haag, The Netherlands.

For the proper interpretation of this figure, however, it must be kept in mind that together with the increased use of certified seed other factors and cultural practices are also improved. For example the use of fertilisers and pest control are usually greater in areas where a high proportion of certified seed is used. Nevertheless the relationship is valid.

accumulative pathogen infection
In the developed world, the quality of the planting material is often limiting. This may especially be the case in horticultural crops that are vegetatively propagated and are thus prone to accumulative pathogen infection.

∏   Think for a moment, what do you think accumulative pathogen infection means?

ramet
ortet
It means that repeated vegetative propagation from field grown plants allows pathogens to multiply and spread through all the propagules (ramets) from a single infected 'mother plant' (ortet). NB we will define ramets and ortets later.

4.2 The potato plant (*Solanum tuberosum* L.)

Since the emphasis of this chapter will be on the potato, it is useful to give you some information about this plant.

The potato plant is the most important non-cereal food crop in the world and is widely grown in temperate, subtropical and tropical regions. It is an annual plant, about 30 - 100 cm tall. Many methods are available for its propagation (Figure 4.2).

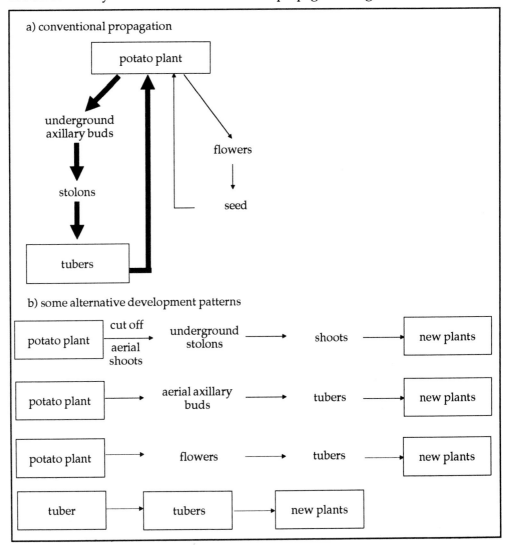

a) conventional propagation

potato plant

underground axillary buds

stolons

tubers

flowers

seed

b) some alternative development patterns

potato plant → cut off aerial shoots → underground stolons → shoots → new plants

potato plant → aerial axillary buds → tubers → new plants

potato plant → flowers → tubers → new plants

tuber → tubers → new plants

Figure 4.2 The potato plant and its propagation.

The traditional (vegetative) mode of propagation is to plant tubers (seed tubers or mother tubers). The buds on these tubers (in the eyes) sprout and develop shoots. The axillary buds at the underground nodes of the shoots form stolons (ie horizontally growing shoots with rudimentary leaves). The tips of these stolons or of the stolon branches may swell if the conditions for this process of tuberisation are favourable. These swellings can grow into new tubers thus forming the new generation. The potato plant may also be propagated generatively by true seed (TPS: true potato seed).

alternative development patterns

Although this is the usual form of development, there can be many variations in the pattern of organ development. The potato plant is very flexible in this respect. Excision of the shoots at soil level may cause the stolons to turn into upward growing shoots, and above ground buds (eg those in the axils of the leaves) may form aerial tubers. It is even possible to form a new generation of tubers directly on the mother tubers or even flowers! This enormous flexibility in the development of the plant is useful for propagation purposes.

As a vegetatively propagated crop, the potato is prone to accumulative infection by:

- bacteria (such as *Erwinia spp.*);

- fungi (such as *Rhizoctonia solani*);

- viruses (such as the potato leaf roll virus);

- viroids (such as the potato spindle tuber viroid).

degeneration effects on yield

This process of accumulating pathogens is called degeneration. Infections with these pathogens may result in dramatic effects on the yield and quality. Moreover these pathogens affect the international distribution of seed and of germplasm. No-one wants to accept diseased material. The risks of accumulation of pathogens are much smaller when true seed is used. The growth vigour of plants from true seed and the quality of the progeny tubers, however, are much lower than of plants from 'mother' tubers. Therefore, it remains attractive to grow potato crops using seed tubers as propagules. (You may prefer the terms 'ortet' and 'ramet' to the use of mother and daughter plants. These latter terms denote femaleness and imply sexual reproduction. Ortet {Latin 'ortus' = origin} and ramet {Latin 'ramus' = branch} are old established botanical terms for the plant from which a clone is derived and for an individual member of a clone).

ortet

ramet

pathogen eradication

Because of the large effect of the use of viable, healthy seed tubers on yield, the high costs of healthy material when it must be bought on the international market, the large chances of reinfection because of the large number of multiplications that take place in the field and the complexity of the breeding of this tetraploid crop, interest in the applicability of the new tissue culture techniques in potato have been very large compared to other food crops. The application of tissue culture as a technique for pathogen eradication and for rapid multiplication has already become widespread in both developed and developing countries and is still gaining in importance.

rapid multiplication

4.3 Overview of the possibilities and methods of plant tissue culture

The current and future applications of plant tissue culture have been already listed in Table 4.1. Here we describe and discuss them in more detail.

4.3.1 Producing disease free propagation material

Many crops are internally infected with pathogens. These pathogens can be transferred to the next generation. Disease free material can often be obtained by isolating meristems (growing areas of the plant) and regenerating plantlets from them. When reinfection can be prevented, it is very easy to produce large amounts of disease free material by subsequent micropropagation. This technique is practised in, amongst other plants, lily, begonia, strawberry and potato. It is important to realise that micropropagation does not guarantee the elimination of pathogens. All

meristems

SAQ 4.1 A crop can be grown from seed or from vegetative propagules (eg potatoes). Which of the following encourage the use of tissue culture techniques to generate vegetative propagules?

1) A high ploidy state of the plant.

2) The occurrence of a large number of diseases of the plant.

3) Heterogeneity in the products produced by sexually derived seeds.

4) A faster, more vigorous growth of the plant from vegetative propagules compared with seed.

5) High value placed on the propagules.

6) Regulations limiting the international transfer of plants and plant propagules.

micropropagated plants must be carefully screened for freedom from disease organisms before they are released.

Meristems are areas of the plant in which cells divide and grow. In Figure 4.3 we have presented a stylised section through a shoot to show where the apical meristem is found. There is a similar meristem just behind the root cap.

explant

The explant used is sometimes composed of only the apical dome (apical meristem), but usually consists of the apical dome along with one or several sub-adjacent young leaf primordia of the sub-apical meristematic section. When the explant contains more leaf primordia and a piece of the portion of stem beyond the subapical region, it is called a shoot tip. There is clearly an optimal size of the explant: when it is too large, there is an increased chance of presence of the viral particles, mycoplasma, fungi or bacteria. When the explant is too small the rate of survival is much lower.

eradication of viruses

The method is most important for eradication of viruses. Particles of systemic viruses have a low titre or are completely absent in the meristem of infected plants, simply because they have not arrived there yet. Therefore this part of the plant can be used to obtain clean material. Moreover, there seems to be a viral eradication power in the *in vitro* system. The mechanism of this eradication remains unknown. Therefore virus infected stocks may be made disease free by using meristem culture methods. The efficiency of the virus eradication may be increased by choosing the right explant (terminal buds are, for example, better than lateral buds), the right physiological status of the mother plant (ortet) and the right conditions during culturing. Moreover there are two possibilities to increase the rate of success:

• Thermotherapy - the plants from which the explants are to be taken can be exposed to a heat treatment (about 36°C). This treatment will not damage the plant too much, but it has a large effect on the activity of the viruses. In most plants (including the potato) eradicating the virus completely by such a heat treatment is not possible without killing the entire plant, but the treatment helps the meristematic tissue to become virus-free.

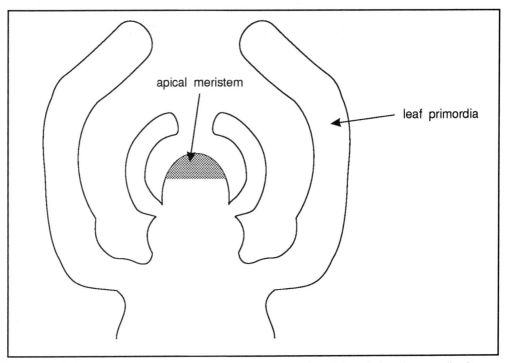

Figure 4.3 A longitudinal section through a shoot to show the position of the apical meristem (stylised).

- Chemotherapy - as was stated earlier, the *in vitro* culture itself has an unknown virus eradicating effect, probably caused by the chemical composition of the medium. There are also many other chemicals of which it is reported that they have a positive effect, such as Virazole (ribavirin) and Benomyl. These compounds are especially useful when larger explants are used.

Note that the thermotherapy is applied to the source plant, whereas the chemotherapy takes place in the *in vitro* phase. We will learn more of this in later chapters.

4.3.2 Rapid multiplication

rapid multiplication and genetic stability

Many crops are vegetatively propagated. Often the rate of multiplication under field conditions is fairly low. Therefore, it takes a lot of time to build up a large population of disease free material or of a new desirable genotype. Rapid multiplication using plant tissue culture techniques offers an elegant solution for this problem as long as these techniques do not affect the genetic stability in the regenerated plantlets. A range of tissues can be used as a source material for rapid multiplication:

- shoot meristems, shoot tips or stem segments with axillary buds;

adventitious shoots and embryos

- tissues that will form adventitious shoots and/or adventitious embryos directly on explants or indirectly via unorganised or partly organised callus. A callus is the name given to a mound of plant cells on solid media.

nodal segments

in vivo techniques

In the case of the potato plant usually meristems, shoot tips or nodal segment cuttings, are used for *in vitro* multiplication. There are however, many possibilities for *in vivo* multiplication as well. Let us examine these each in turn. We begin with the *in vivo* techniques.

If we start with a parent plant, we can choose to use its tubers or seed produced from its flowers, to use an apical cutting, a leaf bud or a shoot tip cutting. Examine Figure 4.4 carefully and follow the fate of each starting material. Then attempt the in-text activity.

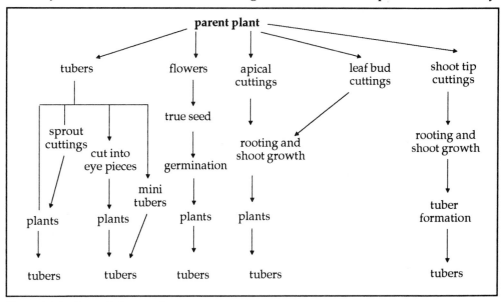

Figure 4.4 *In vivo* multiplication of potatoes.

Π In the *in vivo* procedures for multiplying potatoes illustrated in Figure 4.4, which technique is likely to 1) produce the greatest multiplication of plants, 2) become infected?

The answer to part 1) is not straightforward. Clearly using the 'eyes' from tubers rather than whole tubers could lead to greater multiplication of the number of plants since each tuber has many 'eyes'. Likewise, each shoot has many leaf buds therefore the route using leaf buds rather than apical cuttings should give greater multiplication. Producing seeds will of course produce a very large multiplication factor but could introduce unwanted genetical variability and, in Europe, potato seed ripening and germination and growth are not guaranteed successful.

Although in principle the larger the propagule, the greater the chance of passing on infection, remember that procedures which involve cutting of the plant are likely to lead to infection because of the reduction in the physical integrity of the plant. Infection is of course more likely in *in vivo* situations than when conducted in controlled conditions.

in vitro techniques Now let us turn our attention to the *in vitro* techniques for multiplication. Again we can begin with a parent plant. We can use a variety of tissues or organs. Again follow the fate of each in Figure 4.5.

Figure 4.5 may look complex, but in fact it is rather an oversimplification.

Let us begin with the meristem and shoot tips, by plant regeneration we can grow these to a larger size and use the materials they produce to act as a further source of starting material (eg the shoot tips can be used as a source of nodal cuttings). In the case of potato *in vitro* techniques have also been developed to produce small tubers under sterile

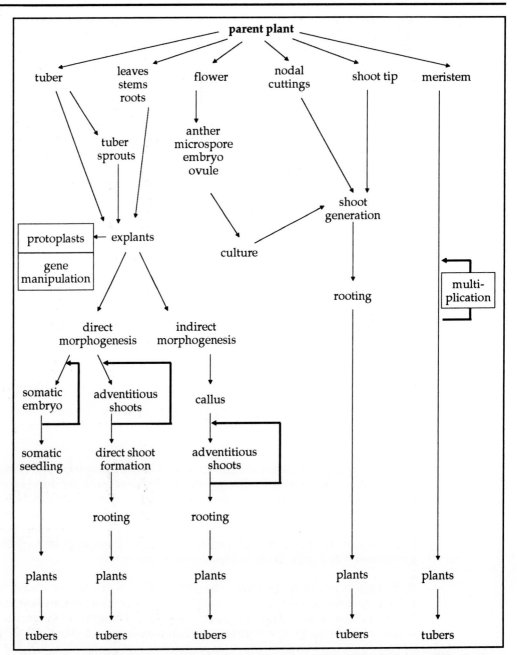

Figure 4.5 *In vitro* multiplication of potatoes. Note recycling loops which enable indefinite multiplication by these *in vitro* manipulations.

conditions. So the final product of the rapid *in vitro* multiplication of potatoes can be *in vitro plants* (plantlets) or *in vitro* tubers.

cell fusion
Note that the explants produced by *in vitro* techniques can be used as a source of single cells for producing protoplasts and for conducting genetic manipulation procedures (cell fusion, recombinant DNA technology). The explants can either directly or indirectly via calluses produce new plantlets (or tubers). The choice of starting materials (eg tubers, flowers etc) depends mainly upon the objective of the *in vitro* cultivation.

recombinant DNA technology

In vitro tubers are sometimes more attractive than plantlets. They are easier to store and to transport. At the same time it is more expensive to produce them. If it is not necessary to store or transport the propagules, there is no need to include a step involving *in vitro* tuberisation. A diagrammatic representation of micropropagation and tuberisation is provided in Figure 4.6.

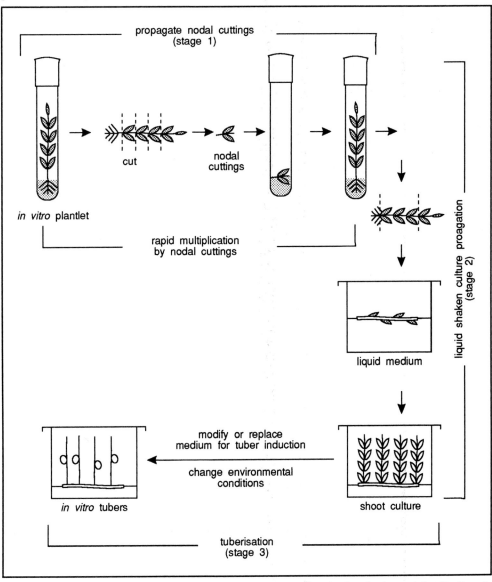

Figure 4.6 *In vitro* micropropagation and tuberisation of potatoes. Note the 3 main phases (after Struik and Lommen 1990, European Association Potato Research proceedings).

So, in the case of potato there are four main techniques of *in vitro* propagation. These include:

- multimeristem culture - this technique is based on bud proliferation on excised shoot tips and subsequent plant regeneration;

- *in vitro* layering - during the culture of stem segments the different axillary buds are able to grow due to lack of apical dominance. The different shoots can be separated and may then develop into complete *in vitro* plantlets;

- propagation from nodal segments. Each individual shoot is cut into pieces with a leaf, a bud and a stem piece. Due to the loss of apical dominance, the bud gets the opportunity to grow out. In some cases it is better to remove the leaf, because senescing large leaves may inhibit the growth of the newly developing shoot;

- *in vitro* tuberisation. This technique actually does not contribute to further multiplication but produces a new type of propagule from the plantlets that has some desirable characteristics that a plantlet does not have.

There are other *in vitro* techniques but these are used in special purposes. We will learn more of these later.

Tuberisation (including aerial tuber formation) and sprouting of tubers are confined to potato and a few other species. The other multiplication techniques can occur in most plant species.

SAQ 4.2

1) Why are potato tubers easier to store and transport than plantlets?

2) Compare Figure 4.4 and 4.5 and make a list of the main differences between *in vivo* and *in vitro* cultivation of potatoes.

Aseptic techniques

surface sterilisation

All tissue culture media contain nutrients and sources of energy (usually sucrose) which provide an ideal environment for the growth of many micro-organisms. If these are not rigorously excluded from the cultures they will quickly overgrow and swamp the plant tissues, which will therefore grow much more slowly. Therefore it is essential that all contaminating micro-organisms are removed from the explant when it is first put into culture. Surface sterilisation is not always sufficient, particularly if there are intercellular spaces in the tissues which can contain internal contaminants. Disinfection of the explant is frequently the most difficult step in many tissue culture operations.

Once a clean culture has been obtained it is essential to maintain aseptic conditions during incubation and to avoid infection with extraneous microbes at the times of transfer to fresh media.

Growing cultures are sometimes protected by sealing the tubes, but this prevents gas exchange. Traditionally, cotton wool plugs, or foil caps are used, which allow gas exchange to take place.

∏ Write down why this is important and what might be the disadvantages of free gas exchange.

The answer is of course that it is important to ensure supply of oxygen for respiration and to allow diffusion of ethylene and CO_2 away from culture. Completely free gas exchange offers the disadvantage that it might allow microbes to enter the culture and there would be water loss from the culture.

Summary of stages in rapid multiplication

A typical sequence of events in the rapid multiplication of a plant is:

- selection and culture of the source plants. Mother (ortet) plants must be tested for pathogens (pathogen elimination may be required), and should be typical of the cultivar;

- establishment of aseptic cultures. Surface sterilisation and transfer of explants to culture;

- multiplication. Multiplication of structures able to produce intact *in vitro* plantlets. This step can be repeated many times;

- preparation for transfer. Plantlets must be rooted;

- transfer to non-sterile conditions. The medium must be removed from the roots and the plants must be placed in a potting medium. Initially the plants must be protected against the natural environment and gradually hardened.

In the case of *in vitro* tuberisation the last two steps are replaced by the following steps:

- establishment of shoot cultures. Shoot cultures are based on a liquid medium, because the flasks must be shaken;

- induction of tuberisation. The medium must be changed (special hormones and a certain amount of sugars are needed) or a small amount of highly concentrated hormone and sucrose solutions must be added. Flasks must be incubated at 18 - 22°C with an eight-hour photoperiod (light intensity of 100 - 500 lux) or in the dark. The temperature and the sucrose content during this step are rather crucial. It is important to note that the physiological processes involved in tuberisation *in vitro* may not be comparable with those *in vivo*;

- after several months small tubers can be harvested. These tubers are dormant. The dormancy is usually somewhat deeper than in the case of normal seed tubers. That means that these small tubers will not sprout unless they are stored for a certain period.

Π To help you remember these schemes, it would be useful to draw for yourself flow diagrams describing the stages.

SAQ 4.3 Make a list of the natural conditions which may be particularly deleterious for the *in vitro* plantlets.

| SAQ 4.4 | Why are *in vitro* tubers more expensive than *in vitro* plantlets? |

Breeding

In vitro techniques can be used in modern breeding programmes, because they may have major advantages over the traditional methods. The most relevant techniques are discussed below. We have listed these advantages and disadvantages in Table 4.2.

Advantages With regard to disease elimination, rapid multiplication and breeding:	- Large numbers of disease free plants can be produced in a short period - Because production takes place under sterile conditions, hardly any losses occur due to pests, diseases, etc - Multiplication can take place in a small space with a strictly controlled environment, therefore whole-year round production is possible - Storage of the propagules is easy - Rapid multiplication is also possible with species that are hard to multiply in other ways - Selection can take place on the level of the individual cell - New genotypes may arise or may be produced
Disadvantages With regard to disease elimination:	- If fungal or bacterial contamination occurs many potential propagules may be lost by internal infection
With regard to rapid multiplication:	- Fairly expensive and highly specialised facilties are required - A large amount of special and skilled labour is required (Possible solution: robotisation) - The propagules are very expensive - Specific methods for rapid multiplication may be needed for each species or even for genotypes within a species - The propagules produced are small, weak and often have a small growth vigour - Rooting is often a problem - During the *in vitro* phase rest may be induced or dormancy may be prolonged - There may be spontaneous mutations or aberrant types - Excretion of toxic compounds may occur
With regard to breeding:	- Genetic stability is often undesirably low - Regeneration is often difficult and variable - Instability and variability of the response of tissue to manipulation *in vitro* - Resistance in vitro may be based on a different mechanism than *in vivo* - Many special problems related to the crop species

Table 4.2 Advantages and disadvantages of plant tissue techniques.

Π Table 4.2 looks a little forbidding, but it is important to remember these features of *in vitro* techniques so read it carefully.

4.3.3 Conservation of germplasm

By *in vitro* conservation, germplasm can be protected against adverse climatic conditions and against pathogens or other threats. *In vitro* conservation is especially important for crops that are vegetatively propagated or have seeds that do not survive normal storage conditions and lose viability quickly. The material is easily available **cryopreservation** when multiplication is wanted. Conservation can be done by cryopreservation (a frozen state) of plantlets, stem tissue, microtubers, pollen, excised meristems, root tips, cell suspensions etc. In the case of potato it is preferable to store excised meristems. The major problem with this type of storage is that material with a high water content is used. Therefore the material must be pre-cultured in such a way that the formation of ice crystals does not harm the explant.

Cryopreservation consists of the following steps:

- culture of source plants;

- excision of the meristems and selection of the healthy meristems;

- pre-culture of the meristems to protect them against the freezing process. Various chemicals that may be added to the culture medium have proved to be useful (eg dimethylsulphoxide, proline and osmotica). The goals of this pre-culture are to produce meristems with small vacuoles and low water content, to influence the form of the ice crystals produced, to protect the meristems against ice damage and to lower the freezing point of the tissue;

- freezing - two strategies are used. The first one is to freeze the material as fast as possible by immersion in liquid nitrogen (temperature -196°C). The other one is slow freezing, but it is difficult to establish the optimum freezing rate. Some systems use a slow cool down (1 - 2°C per min) to about -25°C, followed by immersion in liquid nitrogen;

- storage of the material;

- thawing of the material. This is the most critical stage. Rapid thawing (eg in water of 35 - 40°C) is usually adopted to avoid the damaging effect of ice crystal growth, but many explants die at this stage if it is not correctly done;

- recovery. The solution of cryoprotectants must be washed out and the culture must be nursed to recover from the cryopreservation;

- viability test to determine whether the meristems are viable (eg by judging their growth or their ability to turn green in light).

Π You may find it useful to draw a flow diagram of the stages used in cryopreservation.

Factors influencing the success of cryopreservation are:

- source of the meristem;

- genotype;

- pre-culture conditions;

- the freezing technique;

- the thawing technique;

- the culture technique after the preservation.

minimal growth
storage

As an alternative to cryopreservation, it is also possible to store plant material at a temperature where growth can take place (the so called slow growth technique or minimal growth storage). The plant material can be kept at a temperature of 2 - 4°C with a low light intensity.

growth
retardants

In addition to the cultural conditions the medium may be changed. Growth retardants such as chloromequat, maleic hydrazide, daminozide, osmotica and others can be applied. It is also important to reduce the sugar content of the medium. In the case of potato, tuberisation often takes place during slow growth storage, especially under short day conditions. Tuberisation increases the survival rate.

SAQ 4.5

Make a list of the reasons why you think meristems are preferred for cryopreservation. (Try to think of four or five reasons).

∏ See if you can write down the main difference between rapid and slow freezing.

The main difference is that rapid cooling results in intracellular freezing. This is so fast that little dehydration occurs. Because freezing is fast many small intracellular crystals are formed. If large crystals were formed, then these would disrupt intracellular structures. With slow cooling, extracellular freezing occurs first. This is followed by cytoplasmic dehydration. This reduces the water content of the cells and thus reduces the chance of large, disruptive crystals forming. The rate of cooling is, of course, critical.

4.3.4 Embryo rescue

zygotic
embryos

In some crossings between parents of different species, zygotic embryos (ie embryos produced by the union of two gametes) might be formed that are aborted or are not viable when kept on the plant. This is usually due to endosperm failure as a result of genetic incompatibiltiy. It is sometimes possible to regenerate the immature embryos in an *in vitro* system. This is only useful in those sexual crosses where fertilisation does take place, but embryos fail to develop on the plant. For example, it facilitates the incorporation of genetic material of wild *Solanum* species into normal potato cultivars. In this way a large extra gene pool becomes available, which has already proved to be very useful in breeding for resistance against pests and diseases.

4.3.5 Microspores, anther and ovule cultures

Cultures of microspores (ie small, immature pollen grains), anthers or ovules can be used to obtain haploid plants from diploid genotypes. The number of chromosomes can then be doubled again by chemical treatment. This then homozygous material (ie having identical alleles for any one gene) can be very useful in breeding programmes (see Section 3.7.1). Anther culture as a device for producing haploid cell lines is of increasing importance. Practical details of the processes involved in producing such haploid cell lines are provided in the BIOTOL text 'In vitro Cultivation of Plant Cells'. Although of importance, there are some important limitations in this approach and anther cultures have not been successfully produced from all plants. In the case of tetraploids, such as potato, the matter is more complex. Regeneration of plantlets from these haploid plant parts is not a general trait; it is even highly dependent on genotype. Efficient production of plantlets from these explants can only take place when the growth of the source plant, the sampling of different genotypes, the stage of physiological development, the conditions for pre-treatment, and the composition and physical structure of the growing medium or media are optimised.

∏ Draw a diagram showing the genetic basis for production of homozygous diploid progeny from a heterozygous parent, using microspore culture. Can you explain the advantage of using haploid samples?

In heterozygous genotypes the gametes giving rise to the haploids will show a wide range of recombination. This will result in a wide segregation of genotypes in the resultant diploids. Many will contain lethal recessives and recovery will be low. Plants which do develop will be very variable, and large populations will need to be screened to find the ideal genetic combination. The advantage of using homozygous diploids derived from haploid plants is that the greater the homozygosity of the parent, the less likelihood of recombination and the better the chance of recovering viable plants. Also the products will be homogenous.

4.3.6 Somatic embryogenesis and organogenesis

We have already described that it is possible to regenerate plants from various explants by producing somatic embryos or adventitious shoots. Let us explain these processes a little more fully.

somatic embryos and embryogenesis

A somatic embryo is an independent bipolar structure, which is not physically attached to the tissue of origin and is produced from a somatic cell (ie a cell that is non-reproductive and thus contains two sets of chromosomes). Compare this with a zygote embryo which is produced by sexual fusion (ie the normal product of sexual reproduction). The production of embryo like structures from somatic cells is called somatic embryogenesis. Somatic embryogenesis can occur in two different ways, indirect (after some type of callus culture) or direct (ie without an intervening phase of callus growth). Both types can be used for rapid multiplication. Indirect somatic embryogenesis may be especially attractive as long as the genetic stability is not affected (see below), because the rate of multiplication is very high, the amount of space needed is very low and the manipulation of the cultures is easy. Moreover, the production is well synchronised, making mechanisation possible. Therefore the technique has a great potential. Somatic embryogenesis is mostly used in breeding, rather than for producing a large number of plants for growers.

Π   Let us see if we can devise a protocol for ourselves. Using the stages listed below
     put them in an order that would make a logical scheme.

A - culture on a special solid medium containing an auxin to induce embryo induction.

B - sieve suspension to obtain embryos of uniform size.

C - culture cells as a suspension or a callus.

D - transfer material to a liquid medium not containing auxin.

E - take an explant from a desired source.

We have made this quite simple by confining ourselves to five main stages. Think of
what we will start with and what we hope to achieve. The first stage is to take an explant
E). We might surface sterilise this before we attempt to grow it as a cell suspension
culture or as callus C). We would have at this stage a large number of somatic cells. Next
we would need to induce embryo formation. This is usually achieved by culturing the
cells/callus on solid medium containing auxin A). Subsequently we need to remove or
reduce the auxin to permit embryo development. This is done by culturing the embryos
in a medium devoid of auxin D). Finally we can collect the embryos and grade them
according to size and development B). The order of your stages should therefore have
been ECADB.

organogenesis   Organogenesis can be defined as the transformation of a single cell, callus or tissues into
organ like structures. In the case of formation of adventitious shoots on an explant, we
can discriminate between the following steps:

• dedifferentiation of certain cells of the explant, followed by cell division and callus
  growth;

• redifferentiation of callus cells;

• regeneration and development of plantlets.

Also in this case, the regeneration involves the use of different media, each designed to
perform a specific morphogenetic task.

somatic   Plants can be produced from somatic embryos by 'germinating' them on a culture
embryos as   medium. As soon as they have reached a suitable size and are hardened, they can be
artificial seeds   transferred to a potting medium. In some cases it is desirable to produce 'artificial seeds'
from these somatic embryos. This 'seed' can be encapsulated or coated. Another
possibility is to embed them in a protective gel, containing nutrients, pesticides and
growth regulators. The mix is then planted. This system is called 'fluid drilling' of
somatic embryos.

The absence of a protective seed coat, danger of desiccation of the embryos (which have
a very high water content), shortage of reserves such as starch and proteins and lack of
autotrophism, and the difficulty of obtaining uniform populations are limiting the
agricultural exploitation of these somatic embryos.

| SAQ 4.6 | Why might the production of artificial seeds from somatic embryos with large seeded species such as legumes be especially difficult? |

4.3.7 Mutation and somaclonal variation

somaclonal variation

When plants are regenerated from calli (whether regeneration occurs through somatic embryogenesis or by adventitious shoot formation), it often happens that a new genotype arises. This phenomenon is called somaclonal variation and occurs in many plant species. There are many possible causes, one of which is the naturally occurring variation within a plant, but it is enhanced by the artificial conditions during the tissue culture. The potato is very prone to this type of variation. For example, a large variation in growth habit, tuber shape and size, tuber skin colour, flower characteristics and photoperiod requirements was observed among a series of somaclones of the cultivar Russet Burbank. Also differences in sensitivity to diseases were observed.

The mutation frequency can be increasd by mutagens (chemical or ionising irradiation). The method is interesting as mutations occur as single cell events.

chimera

aneuploids

euploid

In vivo mutation will give rise to a chimera (ie a plant with tissues of different genotypes). Potato plantlets regenerated from somatic cells often do not have the original number of chromosomes. This may occur in 10% of the regenerants. When a plant has gained or lost chromosomes it is called an aneuploid. In the case of regeneration from protoplasts (see below) the proportion of aneuploid plants may be as high as 95%, although this figure can be reduced by manipulating the culture conditions. In addition to changes in chromosome number also translocations and deletions of chromosome parts can be observed. These changes are rather dramatic. Considerable variation can also be induced without changing the number of chromosomes. In these euploid plants, small mutations have taken place and they are often more useful. More information on the use of somatic variation is provided in Chapters 5 and 6.

The use of somaclonal variation has become popular: it may lead to new, useful mutations, which would otherwise never be obtained. Moreover, mutations are very frequent. Yet somaclonal variation should not be overestimated. For example for the potato crop:

- somaclonal variation is probably superfluous: there is a wealth of genetic variation available;

- induction of somaclonal variation is at random and thus very inefficient;

- the use of somaclonal variation in the potato is very time consuming and thus unattractive;

- most products of somaclonal variation are worthless.

Moreover, not all characteristics can change and not all changes are stable. Yet some interesting mutations have been incorporated through this technique, eg mutants with a starch composition that differs from the normal potato plant and mutants with resistance against certain pathogens.

Usually, however, it is desirable to reduce the somaclonal variation especially in germplasm conservation and rapid multiplication. Even when genetic variability is desired, its origin should be better studied before full use of this phenomenon can be made.

Somaclonal variation is influenced by many factors, such as:

- the source plant (species, genotype and ploidy level);

- the origin of the explant;

- the tissue culture procedures employed and the time in culture;

- the composition of the medium (especially the hormones) and the culture conditions;

- the use of mutagens.

SAQ 4.7

A collection of plants, produced by somatic embryogenesis displayed the following characteristics. 100 of the plants produced red flowers, 5 produced yellow flowers. 95 of the red flowered plants were tall, 5 were dwarf. The yellow flowered plants were tall. The cells of the tall plants contained 24 chromosomes, those of the dwarf plants contained 25 chromosomes. The cells of the original parent plant from which the embryos were derived contained 24 chromosomes. One plant produced both red and yellow flowers on separate branches.

Use the list of terms below to complete the following statements.

1) The dwarf plants which appeared to have gained an extra chromosome can be regarded as [   ].

2) The tall red and yellow plants can be described as [   ] plants.

3) The plant producing both red and yellow flowers is probably a [   ] plant.

Word list: **chimeric, polyploidal, euploidal, aneuploidal, haploid.**

4.3.8 Selection *in vitro*

in vitro selection

An increased rate and level of selection is often attempted by using *in vitro* selection systems. In such cases desired mutants are selected from numerous individuals. This can be done much more efficiently in cell cultures where the rate of mutation is higher and the number of individuals much larger than is possible with plants *in vivo*.

herbicide resistance

Examples are selection for resistance against certain herbicides (which then can be used very efficiently in that crop), tolerance against salt, heavy metals, toxins and biotic stresses, quality characteristics, etc. It is not always clear whether the mechanisms of resistance are similar for the *in vitro* system as for the whole plant.

The design of the best selection system, however, is very difficult as long as one does not understand the biochemical basis of the desired mutant. Here the fact that we still know so little about the functioning of the entire plant strongly acts against us. Moreover it is not certain whether the characteristic that is shown in the cell or tissue culture will also be expressed after plant regeneration. More information about the selection of mutants is provided in Chapter 5.

4.3.9 Culture and fusion of protoplasts

somatic
hybridisation

pomato

topato

Protoplasts are somatic cells from which the cell walls are removed by enzymatic digestion. By fusing two of such cells from two different species, it is possible to combine the genetic material of species that can normally not be crossed. By this somatic hybridisation a new 'species' can be created, in which desirable traits of the first two species are combined. A well known example is the fusion of protoplasts of potato and tomato resulting in somatic hybrids that are called pomatoes or topatoes.

∏ Use a full sheet of paper to draw out a flow diagram of these stages as you read through the rest of this section.

The procedure consists of several steps.

Growth of source cells, tissues or plants

importance of
culture
conditions

It is important for later steps (especially the regeneration step), that the source material has been produced under well defined and controlled conditions of temperature, light intensity, photoperiod, and nutrition. The response to the culture conditions depends on the physiological stage of the plant, tissue or cells. There is also a need for preconditioning the source material in order to get a maximum production of the protoplasts. There are many reports that the number of protoplasts produced depends on the type of tissue of the source. Because of this *in vitro* plantlets or cell suspensions are often used as source material.

Isolation of protoplasts

use of wall
degrading
enzymes

This is usually done by treating tissues or cells with a mixture of cell wall degrading enzymes (such as different types of cellulase, pectinase, pectolyase) in an osmotically stable solution which is hypertonic to the plant tissues.

∏ What would happen to the protoplasts without the presence of an osmoticum?

They would of course imbibe water, swell and probably burst (lyse). The success of this step depends on the type of explant, the mixture of enzymes and the procedure used. With the potato plant, mesophyll cells are often used, but it is also possible to use tuber tissue or stem tissue. It is important to remember that the physiology of the tissue changes rapidly as soon as it is detached from the plant. This means that the tissue must be protected against degradation, or be used shortly after excision from the plant.

plasmolysation

Sometimes special treatments are needed to enhance the penetration of the degrading enzymes into the tissue. Moreover, a step of plasmolysation (ie the separation of the protoplast from the cell wall by loss of water in the cell through bringing the cell into contact with a hypertonic solution) before the digestion of the tissue may improve the frequency of good protoplasts.

The physical conditions during the cell wall digestion (such as temperature and rate of rotation of the shaker) may have some effect on the number of protoplasts obtained.

This step of protoplast production is however usually not very difficult and many adequate techniques have already been developed.

Purification of the protoplasts

In the mixture of cells obtained there are many cell fragments, broken cells, cells with incomplete cell wall digestion, cell clumps and vascular tissue. From this mixture the good protoplasts must be separated. This can be done by a combination of filtration and centrifugation. Instead of, or in combination with, centrifugation one can also use a sucrose floatation technique. In this system a density gradient is constructed in a centrifuge tube using a series of different sucrose concentrations. When the suspension of protoplasts and debris is centrifuged, the intact protoplasts form a layer at the appropriate sucrose density and the debris forms layers at other densities, thus separating them.

Hybridisation of the protoplasts

fusion

electrofusion

This step involves the actual fusion of the protoplasts and must be done before wall regeneration occurs. It can be done in two ways: chemically induced fusion (eg by osmotica such as polyethyleneglycol and dimethylsulphoxide) or electrofusion. The latter involves two steps. In the first step, protoplasts are placed in a medium with a low conductivity and a high frequency alternating current field is applied. The protoplasts then start to act as dipoles, migrating to regions of higher field intensity. During this movement they come in contact with other protoplasts and form chains. As soon as these chains are formed, the second step starts. A direct electric pulse is applied, causing reversible breakdown of membranes. Contacting membranes may then fuse. The alternating current field is then briefly reapplied to make sure that close membrane contact is maintained as fusion begins. After that the electric field is reduced to zero. There are many minor variations on this general process. We will examine electrofusion in more detail in a later chapter. Very high proportions of aligned protoplasts may indeed fuse.

Proliferation of protoplasts

Protoplasts undergo considerable cytological changes, for example, cell wall regeneration, DNA synthesis, mitosis (nuclear division without reduction of the number of chromosomes) and cytokinesis (the separation of cytoplasm that follows nuclear division). These are necessary steps in the dedifferentiation of the cells and the subsequent shoot-bud morphogenesis. After dedifferentiation the cell division must be as efficient as possible. This phase requires special media, containing growth regulators, minerals, growth factors and sugars etc. The culture requirements are species and even genotype specific. The cell density is also crucial: too many cells per unit of volume would be inhibitory for cell division; not enough cells will also inhibit initial growth.

Regeneration of protoplasts

This process is a crucial step and consists of the induction of shoot morphogenesis, elongation and rooting of the shoots. After the cell wall regeneration and the initial cell divisions, the protoplast calli are formed. They are further cultured in a multi-step medium sequence to stimulate callus growth, shoot initiation, shoot elongation and rooting (cf embryogenesis and organogenesis). Methods may vary greatly.

Selection of desirable hybrids

Desirable hybrids can be identified more easily by the use of genetic markers. These markers should cause a minimum disruption of parental and hybrid cell genotypes, should enable selection early in the regeneration process, should allow large numbers to be screened within a reasonable time and should be unambiguous. The best markers are those that make it possible to kill the undesired protoplasts or inhibit their growth.

Other possibilities are biochemical markers (eg inducing the production of secondary metabolites), and those inducing gross phenotypic differences (eg chlorophyll defects).

In Chapter 6 we will discuss the somatic fusion technique in more detail.

SAQ 4.8

1) Select the best type of tissue to use for protoplast preparation from the list provided:

   a) meristems;

   b) unopened flower buds;

   c) leaf mesophyll;

   d) thick walled seeds.

2) Which of the following are shoot meristems used for?

   a) micropropagation;

   b) anther culture;

   c) cryopreservation;

   d) protoplast preparation.

3) Select the correct term to complete the following statement.

   Incubation of a plant cell in a strong sugar or salt solution will cause the protoplasm to shrink away from the cell wall. This process is called:

   a) hybridisation;

   b) hydrolysis;

   c) cytokinesis;

   d) plasmolysis;

   e) fusion.

4.3.10 Transformation

With the most recent techniques it is possible to bring in new, well defined genetic material into a selected cell. This is called transformation. In this way a chosen characteristic may be built into a crop. It may be done by using:

- *Agrobacterium*, DNA viruses or a combination of these two (indirect gene transfer);

- directly by micro-injection of DNA into nuclei;

- fusion of protoplasts with DNA containing liposomes (ie fatty or oily globules in the cytoplasm of a cell);

- gene transfer into protoplasts mediated by chemicals or electric fields;

- macro-injection;

- shotgun transformation.

Other techniques are also being developed, but (until now) without much success. Unfortunately we cannot go into detail on these interesting new developments at this stage, but Chapter 8 will tell you more about them.

SAQ 4.9	Draw a flow diagram to show how a hybrid plant may be produced by use of protoplast fusion. Begin with two separate plants and finish with a new hybrid plant.

4.4 Production of chemical compounds by cultured cells

Many plants contain compounds that are interesting for mankind. Often these compounds, such as protein and starch occur in large quantities in many species. In other cases, these compounds are restricted to certain species and quite often also confined to certain plant parts of these particular species. Examples of compounds with restricted distribution are certain fatty acids, alkaloids, polyphenols etc.

The value of these compounds is not always positive: they may be toxic or reduce uptake of feed. In these instances, however, they may be valuable for the plant as protectants against diseases, pests or animals. Sometimes, however, these compounds have a high commercial value.

stereo-specific products

The commercially interesting compounds are often compounds that are difficult to synthesise. Synthetic methods often produce unwanted stereo-isomers and contain impurities from side reactions and traces of reactants. Biologically useful molecules are usually specific stereo-isomers. The plants ability to produce these compounds in considerable quantities and in a very pure state can be used by man by growing the crops, harvesting the relevant plant parts and extracting the compounds from these parts. The efficiency, however, is not very great. Therefore there is an interest in the production of these so-called secondary metabolites by culturing plant cells. The problem is that the levels of the interesting compounds are then often much lower than in the plants themselves. Yet the advantages are obvious: continuous production, independent of growing season, facilitated sampling of the material, reusability of biomass, etc.

secondary metabolites

The commercial exploitation of these techniques using plant cells is realised for shikonine (a pigment), berberine (a drug) and ginseng (also a drug). For other substances (eg an agrochemical against nematodes from *Tagetes spp.*), commercial cell culture techniques are under investigation. Micro-organisms, however, are already used on a large scale in the processing industry. Potential application of the techniques using plant cells can be classified into five groups: drugs, flavours, perfumes, pigments and agrochemicals (we will learn more of some of these in Chapter 5).

4.5 Closing remarks with regard to the methods and possibilities which arise from *in vitro* plant culture

The range of sophisticated techniques in tissue culture is broad. Some of the techniques are already having a direct and significant effect in many crop species. Other techniques are still in the stage of development. How great the potential of these techniques is has still to be proved.

Although the potential of the *in vitro* techniques is enormous and will revolutionise (or has already revolutionised) plant breeding and multiplication techniques, there are also problems. Table 4.2 contains some major advantages and disadvantages of plant tissue culture techniques.

SAQ 4.10	Plant regeneration is considered to be the major problem in the use of plant tissue techniques. There are three routes. One method is organogenesis. Name the two other methods through which plant regeneration can be accomplished. Which one of the three is the best for rapid multiplication? Which one gives the highest rates of multiplication?

4.6 Application of *in vitro* techniques in the multiplication of the potato plant (*Solanum tuberosum* L.)

Figure 4.7 contains a diagram that summarises the principal methods of plant tissue culture of potato. In this scheme the use of potato cell cultures to produce interesting compounds is not presented.

∏ Figure 4.7 is quite complex, examine it carefully. Which two methods for long term storage (other than tuberisation) are illustrated?

The answers we hoped you would find were cryopreservation and sub-culture using low temperature or growth retardants.

4.6.1 General characteristics of seed programmes

In potato seed programmes the following goals are very important:

- to multiply as fast as possible and as cheaply as possible, healthy individuals from desired genotypes;

- to keep the propagation material free from diseases.

After eradication of viral and other diseases, the new healthy material can be multiplied several times before it is transferred to the open field. Thereafter it is multiplied in the open field several times. During these multiplications, there is a risk of infection, depending on the conditions. Reducing the number of multiplications in the field usually improves the standard of seed. Such a reduction can be obtained by increasing the quantity of the nuclear stock and increasing the multiplication rate. The first

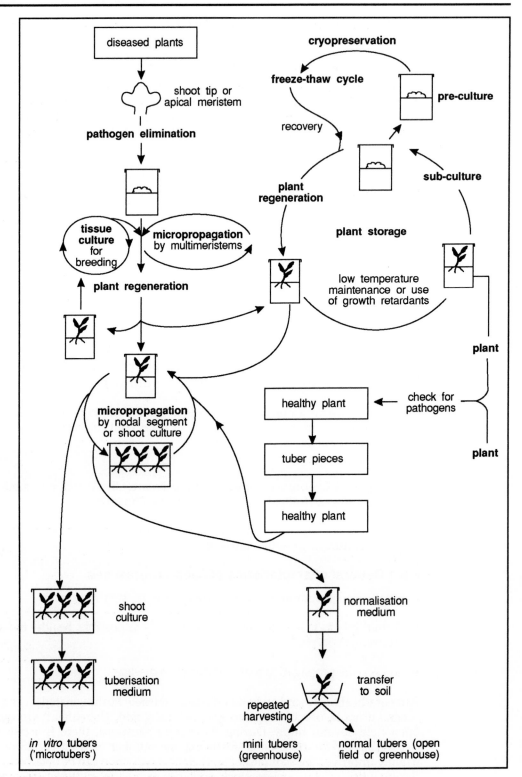

Figure 4.7 The application of *in vitro* techniques to potato.

possibility is the more feasible one since the new plant tissue techniques have become available.

4.6.2 Traditional methods of propagation

clonal selection

As stated before, the traditional way of propagating the potato crop is by producing a new generation of tubers. In many countries healthy seed is produced by repeatedly propagating a sample of tubers that have been proved to be completely free of pathogens. This system is called clonal selection and the principle is shown in Figure 4.8.

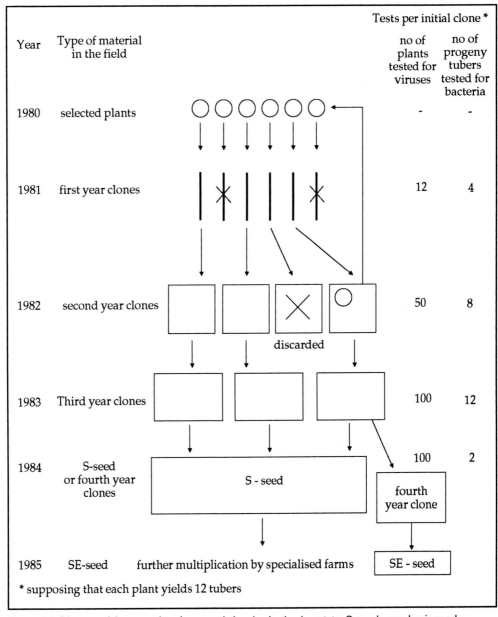

Figure 4.8 Diagram of the procedure for normal clonal selection in potato. S-seed = pre-basic seed. SE-seed = basic seed.

Use this figure to follow the description below.

Plants from selected and controlled fields and their neighbours are checked for diseases and then selected as source plants. These selected plants are multiplied in the field under strict regulations, with continuous control of pests and diseases and with frequent inspection and testing of occurrence of disease infections. Entire clones are discarded when infection is observed. Part of the inspection is done on special centralised fields. Not only viral infection, but also infection by bacteria, fungi (*Phoma*) and other diseases are tested. Moreover, it is checked whether the clone is 'typical' for the variety and pure (ie uniform). An important aspect of the production of healthy seed is the killing of the haulm before a new generation of winged aphids has developed. Aphids are able to transfer viruses from one plant to another (not necessarily of the same species) when they are mobile. The rate of development of the aphids strongly depends on temperature and other factors. It is impossible to protect the plants against these virus infections carried by aphids with the use of insecticides, because the aphids stay alive after the uptake of the insecticide for at least 6 - 8 hours.

Therefore the development of the population of winged aphids is recorded. This information is used to decide a date of mandatory haulm killing. At such a date the tuber growth is still in progress; the crop is actually killed at an immature stage. The rate of development of the crop is therefore the major determinant for the obtainable yield. This rate can be influenced by the pretreatment of the mother tubers or by cultural practice.

multiplication farmers

The progeny of the third year clone may be sold to specialised multiplication farmers or may be multiplied once more by the farmer who carried out the clonal selection. The multiplication farms can use the seed to multiply it over several years, but with each year of multiplication the quality declines and the stock is downgraded at least one class.

It may be clear from Figure 4.8 that the system of clonal selection has a slow rate of multiplication and is expensive and laborious.

SAQ 4.11

Seed tubers which are used for the production of a seed crop are usually pre-sprouted before they are planted. Can you think of a reason why? (Think about what happens to these sprouts after the 'seeds' are planted).

4.6.3 Alternative procedures using plant tissue culture

The system of clonal selection described above is very expensive and laborious. It requires intensive control of disease infection, frequent inspection and has a low rate of multiplication. Because of this low rate, it takes many years of field multiplication to build up a population and this results in high risks of infection with bacteria, viruses and fungi. There are more disadvantages of the system: the seed tubers are bulky and require long term storage in a controlled environment after production. Because of these disadvantages, alternative schemes have been developed, which enable the farms to start with a larger nuclear stock or increase the rate of multiplication. These alternative schemes are illustrated in Figure 4.9.

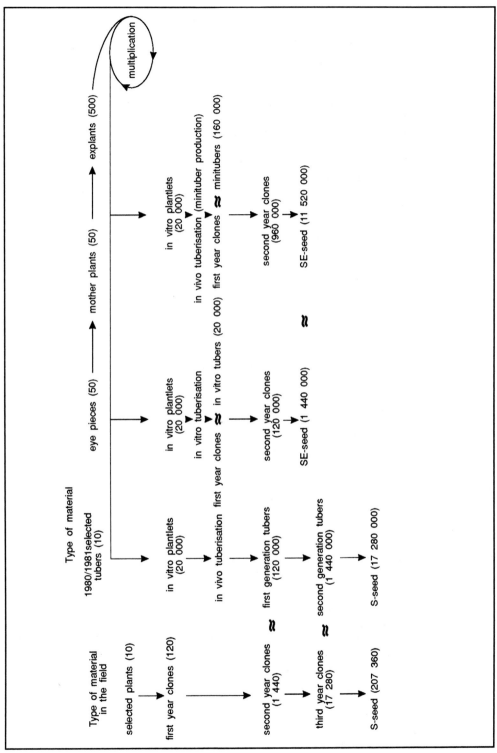

Figure 4.9 Possible (not yet fully authorised) scheme of the use of different types of propagules in the clonal selection of potato. Between brackets is an estimate of the number of individuals. S-seed = pre-basic seed; SE-seed is basic seed.

⊓   Note the rapid expansion of the number of plants in each clone using the newer techniques. Compare this with traditional potato growth. Assume that each seed tuber gives rise to 12 tubers. If we started with 10 tubers, one growing season later we could have 120 tubers. How many would we have 2 seasons later? How many SE-seeds could we have produced by this time using *in vitro* plantlets and *in vivo* tuberisation? (The answers are in Figure 4.9).

Your answer to the first part should have been 17280 (ie 120 x 12 x 12). The answer to the second part is 11 520 000 (follow the right hand column in Figure 4.9).

From selected tubers, cuttings can be produced that are multiplied *in vitro* according to the procedures described earlier. These *in vitro* plantlets are used for:

• the production of *in vitro* tubers;

• transfer to the field for normal production of tubers. This is usually done in screenhouses to protect the plants against aphids;

• transfer to the glasshouse for the production of minitubers.

The progeny tubers of the *in vitro* plantlets are considered as first year clones when they are produced *in vitro* or as second year clones when they are produced *in vivo*. The minitubers, produced under partly controlled conditions of the glasshouses are also considered as first year clones. The rate of multiplication strongly depends on the rate at which the *in vitro* plantlets are built up. This rate varies with genotype. The rate used in the scheme indicated in Figure 4.9 is conservative.

It should be noted that the number of tubers harvested per plant is lower when plants are grown from *in vitro* plantlets, *in vitro* tubers or minitubers than when one uses normal seed tubers. But in all cases the overall rate of multiplication is much faster than for the traditional clonal selection and fewer years of multiplication are required.

The following section discusses the production of minitubers in more detail. Depending on the ratios between the expenses of glasshouse space and labour two techniques can be used:

• 'normal' production of tubers from *in vitro* plantlets;

• 'repeated harvesting' of tubers.

repetitive tuber   The second system involves an early harvest of the plant at which the small tubers that
formation   are already formed and are large enough to be used, are removed. The plantlet is then replanted and forced to produce tubers again. This method can be applied several times. In this way, the plant produces more tubers than usual, although the tuber yield may be lower at each tuberisation.

These alternative schemes lack some of the disadvantages of the traditional methods, but there are some new problems. The alternative schemes are very costly and require large investments and considerable amounts of labour. Moreover, there are some agronomic problems that need to be solved before the system will be feasible:

- the dormancy of the *in vitro* tubers and the minitubers is long;

- the small tubers and the *in vitro* plantlets cannot be planted early in spring; frost is more detrimental to them than for normal seed tubers;

- the slow development of a potato crop grown from very small seed tubers reduces the yield;

- the slow development also has consequences for the susceptibility to pests and diseases. In normal seed crops a certain amount of resistance is built up during the growing season. This mature plant's resistance is absent or weaker in plants grown from very small tubers or from *in vitro* plantlets. That means that haulm killing must occur at an earlier date, causing the growing season to be shortened at both ends.

Figure 4.10 shows the development of potato crops, measured as leaf area index (ratio of leaf area to ground area) against time for a variety of different propagules.

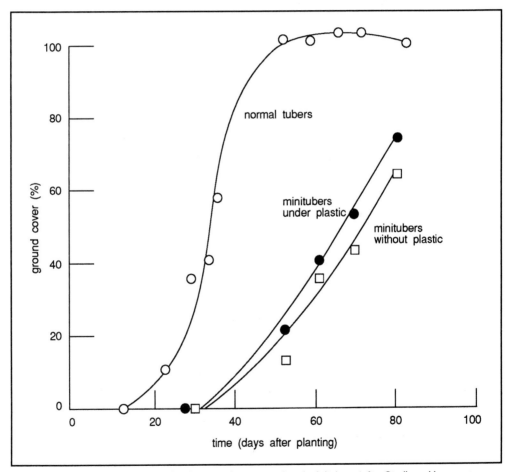

Figure 4.10 Rate of ground cover development from normal and minitubers (after Struik and Lommen, 1990 European Association Potato Research proceedings).

∏   Examine Figure 4.10 carefully. Note the difference in growth rate of plants from
      normal tubers and microtubers with or without plastic. The slower growth rate
      of potatoes from small seed also increases the negative effect of spring frosts, the
      problems of weed control and the occurrence of physiological disorders such as
      'second growth' (ie the production of secondary tubers on primary tubers after a
      period of conditions that are unfavourable for tuber formation). Note the effects
      of cultivation under plastic covers. The rate of canopy development can be slightly
      increased by covering the soil with transparent plastic.

It is very important to develop cultivation techniques to enhance early growth.

| SAQ 4.12 | Why is the control of weeds and physiological disorders more difficult when the early growth is slow? |

| SAQ 4.13 | What kind of cultivation techniques could be used to reduce the problems associated with the slow early development? |

Summary and objectives

This chapter has examined the use of plant tissue culture in providing new ways of cloning plant materials, providing plant stocks of assured quality and to provide material suitable for genetic manipulation. Emphasis was placed on potato as an example of the application on *in vitro* cultivation technology.

Now that you have completed this chapter you should be able to:

- list the uses of plant tissue culture techniques and explain their advantages and disadvantages over conventional culture techniques;

- use knowledge of the advantages and limitations of plant tissue culture techniques to identify those circumstances in which *in vitro* cultivation should be the method of choice;

- select appropriate tissues for cryopreservation and for protoplast preparation;

- draw outline schemes for the production and selection of hybrid plants derived from protoplast fusion;

- define the terms somatic embryogenesis, zygotic embryogenesis and organogenesis and explain the use of each in the production of plantlets from *in vitro* plant cultures;

- explain what is meant by the term somaclonal variation and describe the potential use and problems which arise from this variation;

- use the terms chimera, aneuploid, euploid and embryo rescue;

- compare the multiplication rates available by conventional and *in vitro* culture techniques with particular reference to the potato.

Variation and mutant selection in plant cell and tissue cultures

5.1 Introduction   100

5.2 The nature and origin of variation in tissue culture   100

5.3 The selection of mutants from cell and tissue cultures   108

Summary and objectives   115

Variation and mutant selection in plant cell and tissue cultures

5.1 Introduction

The growth and behaviour of somatic plant cells in a tissue culture system resembles what happens during cell division in a plant growing in a field or greenhouse. In such a plant all cells are generally regarded as genetically identical, although gene expression may differ between cells. Since individual cells can be regenerated into whole plants in many plant species, these genetically uniform cells should give rise to identical plants. Such a group of plants with the same genotype is called a clone. Cultivars of many crop species eg potato, fruit trees and many ornamentals are in fact clones, which are propagated by tubers (eg potato), via cuttings (eg many ornamentals) or by grafting shoots on rootstocks (eg roses and fruit trees). Because multiplication in tissue culture is often faster, this procedure has become an important new technique for vegetative propagation of crop plants. However, problems arise when all cells within an individual are not genetically identical or when genetic changes originate in the tissue culture. The consequence of this is that variation occurs among the plants derived from one genotype via tissue culture. This is undesirable for the growers. This variation however can also be useful for those that want to produce new plant varieties. Plant breeders are always looking for plants which are 'better' than the existing cultivars. The classical approach is to 'create' variation by making crosses and selfings, by looking for (rare) spontaneous mutations or by inducing such mutations. Variation that can be found or

genetic
variations

induced in tissue cultures provides an additional source of genetic variation for plant breeders. The selection of the best genotype among the many variants in an efficient way is the step that follows the induction of variation.

In this chapter we will first discuss the variation that (sometimes) occurs in tissue culture and which is unwanted in vegetative propagation and secondly how one can use this variation to select useful mutants.

5.2 The nature and origin of variation in tissue culture

5.2.1 The different types of variation

physiological
response

Although the somatic cells of one individual are more or less genetically identical, the cells of various tissues (eg roots, leaves, flowers) differ in appearance. This is so because in addition to many 'housekeeping' genes, specific genes are expressed in specific cell types. With this in mind it is not a surprise that tissue culture cells can also show variation because of differences in gene-expression between cells. When this is due to

epigenetic
changes

external factors (eg plant hormones in the tissue culture medium) which cause a change in the appearance of the cell culture (eg shoots appear on a cytokinin containing medium) and this change is reversed when one removes the inducing factor, this change is called a physiological response. However one sometimes also observes aberrant cells where the cell-phenotype is not reversed by a change in conditions. For example a cell within a callus becomes pigmented. This pigmentation is also present in the cells derived from this cell by cell division. Such a change in the cell phenotype does not have to be due to a mutation but can be caused by the switching-on of previously inactive pigmentation producing genes in these cells. Such stable changes in gene expression, which are transmitted to the daughter cells, are called epigenetic changes.

reversibility of
epigenic
changes

After fusion of two gametes, or cells from tissue culture that have dedifferentiated (eg during the formation of callus tissue) the genes characteristic for a certain state of differentiation are switched off. Similarly epigenetic changes are mostly reversed after dramatic changes in the state of differentiation of the cells (eg by the regeneration of plants from such cells and by meiosis). This is in contrast to genetic changes, which are due to changes in the genetic material of the cell and which remain the same in the cell-lineage. A number of criteria that can be used to distinguish epigenetic from genetic changes are listed in Table 5.1. None of these characteristics is as such diagnostic for one or the other phenomenon with the exception of the transmission via meiosis.

Epigenetic changes	Genetic changes
- Often inducible with specific treatments	- Random
- Relatively frequent	- Relatively rare
- Reversible (eg via plant regeneration)	- Not reversible
- Not transmitted via meiosis	- Transmittable via meiosis

Table 5.1 Characteristics of epigenetic and genetic changes.

definition of a
variant

Since the isolation of a cell line that differs in some phenotypic aspect(s) from the original culture does not mean that one is dealing with a change in the genetic material (a mutation), one should call such a cell-line a variant. It can be described as a mutant when one has shown that the phenotypic change is heritable because it is transmitted to the sexual progeny of the plant regenerated from this cell-line. There are cases (eg some potato cultivars) where the plants are completely sterile and therefore such a progeny test cannot be done. Here the stable vegetative propagation of a variant phenotype is often also considered to be a good indication that one has a mutant. Also chromosomal changes that can be identified with a microscope do not need a progeny test to establish that a genetic change has occurred.

SAQ 5.1

A collection of somatic embryos have been produced from a single callus. These embryos were weaned onto soil and eventually planted in a field. A small proportion (1%) of the plants were shown to produce defective flowers and could only be propagated by shoot culture. The remaining plants could be propagated via seed production.

1) Should the plants which produce defective flowers be regarded as variants or as mutants?

2) Examination of the cells of the plants producing defective flowers showed they contained 37 chromosomes whereas the cells from normal flowering plants contained 36 chromosomes. What can you conclude from this?

5.2.2 Types of genetic variation

Genetic changes (mutations) can be subdivided into different types depending on the nature and consequences of the mutation that has occurred in comparison with the original genotype (the wild-type).

wild types

chromosomal mutation

When a change is visible in the chromosomes using microscopic techniques, it is called a chromosomal mutation. These can be subdivided into:

- genome mutations - when the number of chromosome sets (genomes) within a cell has been changed resulting in polyploid cells in the case when additional genomes are present;

- chromosome number mutations - when copies of individual chromosomes are lost or additional copies are present. This results in aneuploid cells;

- chromosome structure mutation - where the shape of chromosomes has been altered eg because parts of the chromosomes have been deleted (deletions) or when parts of different chromosomes form a new chromosome (translocations).

Such changes that can be observed under a microscope imply substantial changes at the molecular level. Smaller molecular changes may lead to sub-microscopic modifications in chromosomes. Mutations that only affect individual genes are not microscopically visible and their existence is derived from the effect of the mutation on the phenotype of the cell or plant. The analysis at the DNA level will show such mutations as base pair changes, small deletions etc, and this analysis will also show mutations which are present that have no phenotypic effect in the plant.

extra-chromosal inheritance

All of the types of genetic changes mentioned above affect the genetic information present on the chromosomes which are in the nucleus of the cell. Mutations, however, may also affect the DNA contained within the mitochondria and chloroplasts. Since organelles are only transmitted through the female gamete, mutations in the mitochondrial or chloroplast genome inherit maternally.

5.2.3 Somaclonal variation

somaclonal variation

When mutations, especially of the chromosomal type, can already be observed in tissue culture cells, it is not a surprise that plants regenerated from such cultures are also often different from each other. This variation among regenerated plants was called somaclonal variation by Larkin and Scowcroft in 1981 and this term is now in general use. Some mutations in the cell-culture allow growth as callus but block differentiation and make it impossible for a viable and normal plant to develop from such a culture. Plant regeneration, with its more severe constraints to the functioning of cells as compared to a dedifferentiated callus cell, acts as a kind of sieve for mutations with too severe effects. Thus the genetic variation in a population of plants regenerated from callus cells may be different from the genetic variation present in the callus cell population.

In addition, the fact that many mutations are only expressed in differentiated cells and tissues, implies that many mutations are only observed for the first time in the regenerated plants. For example, one only sees a flower colour mutation on a plant that has flowers. The fact that most plants and plant cells are not haploid means that recessive mutations are not even observed in the primary regenerant but only in their progeny.

There is as yet no explanation of why so many somaclonal variants occur in regenerated plants. Evidence is accumulating for the possibility of mitotic recombination, and for a pseudo meiosis preceding *in vitro* somatic embryogenesis. Just as with classical mutagenesis, recessive mutations of specific genes may be found immediately after the (mutagen/tissue culture) treatment. The phenotypic effects of chromosomal mutations will often show up in the first generation especially in the case of genome mutations and translocations resulting in, for example, sterility.

The genetic consequences and the terminology for the different generations compared with that for classical mutagenesis and sexual crossings are shown in Figure 5.1.

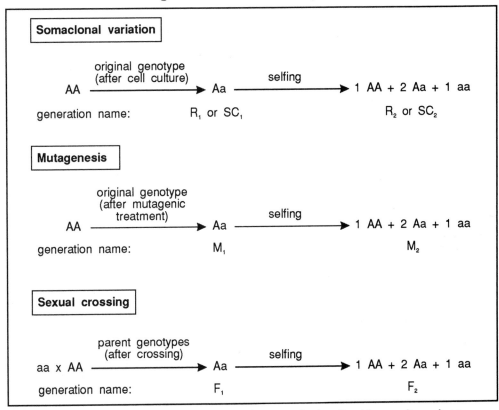

Figure 5.1 Genetic consequences of recessive mutations occurring in cell and tissue culture, after a mutagenic treatment of seeds or plants and after a homozygous mutant has been crossed with a wild type.

Π   Put a circle round the individuals in Figure 5.1 which would show phenotypic expression of the recessive allele. Note our labelling - the gene is called 'A' and of this gene there are two forms (alleles): A (the dominant allele) and a (the recessive allele).

Your answer should be to circle all the aa genotypes found in the R_2 (SC_2), M_2 and F2 generations. All the rest would show the A phenotype since the A allele is dominant.

mutation spectrum   The analysis of many plants regenerated from cell and tissue cultures has confirmed that all of the types of genetic changes that we have described can be found. Before using somaclonal variation for plant breeding, it is important to know if the type of mutations (the mutant spectrum) is different and hopefully 'better' than that obtained with classical mutagenesis. Until now no clear indications for this have been found.

<table>
<tr><td>

SAQ 5.2

</td><td>

Assume that flower colour in a particular plant species is inherited by a single gene.

Five plants (labelled A, B, C, D and E) were derived from a homozygous, red flowered parent (FF) by *in vitro* culture techniques. These five plants were crossed with themselves and the flower colour of the progeny was determined. The results are recorded below.

		Number of progeny	
Plant	Cross	Red flower	Yellow flower
A	AxA	208	0
B	BxB	134	41
C	CxC	143	0
D	DxD	149	0
E	ExE	107	0

1) Determine which of the plants listed in the first column carried recessive (f) flower colour genes.

2) What can you conclude about the stability of the flower colour gene?

3) What would be the result if E was crossed with B?

</td></tr>
</table>

5.2.4 The origin and causes of genetic variation in cell and tissue cultures

chromosome
stability in
tissue culture

A certain basic chromosome number is a characteristic for a plant species. This is an indication that the germline cells and meristems that give rise to the vegetative or sexual progeny of the plant are genetically very stable. However, in many cases differentiated cells can be found within a plant that are polyploid. If such cells are cultivated and then regenerated into 'new' whole plants these of course also will be polyploid. The frequency of chromosomal mutations often increases with the time of tissue culture. This indicates that cell and tissue culture itself leads to these mutations. In several cases it has been observed that chromosomal changes occur rather soon after the tissue culture was started. This suggests that especially the first cell divisions of the (often differentiated) cells, which are activated to start dividing again, are liable to genetic accidents.

To investigate if single-gene changes already exist before the cell culture is started or if these arise in culture, experiments have been performed that used protoplasts. Cells can be regenerated from single protoplasts. Then such cells can be used to produce regenerated plants. Thus in these experiments, protoplasts are prepared from cells derived from explants from plants known to be heterozygous for easily recognised alleles. If the original plant contained genetically identical cells but mutations arose during tissue culture, we would anticipate that different plants produced from a single protoplast may sometimes have different phenotypes. If on the other hand, the original plant was carrying a few mutant cells then when the protoplasts are produced, a few will carry these mutations. Each 'clone' of plants produced from the protoplasts will produce only a single phenotype. In practice these types of experiments show that mutations may arise both in the 'parent' plant and also during tissue culture (see Figure 5.2).

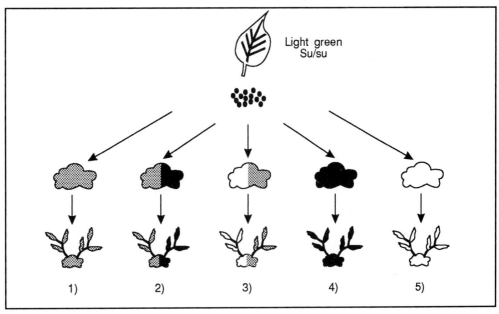

Figure 5.2 Plant regeneration pattern from colonies derived from single protoplasts isolated from light green (Su/su) leaves. 1) All shoots normal, phenotypically parental type regenerants: 2) and 3) heterogenous colonies from which at least one shoot was non-parental, interpreted as cell-culture generated variation: 4) and 5) homogenous colonies from which regenerants are either non-parental dark green 4) or albino 5), interpreted as pre-existing variation.

What is the reason for the genetic instability in plant cell cultures?

We can identify several reasons.

incomplete replication   The doubling of the chromosome number and the origin of aneuploidy is known to be due to irregularities during cell divisions. The segregation of chromatids (the two sister chromosomes that originate from one chromosome by replication), is often not completely normal in cell cultures. Especially, late replicating parts of the chromosomes may not have replicated completely before the chromatids start to separate. This can lead to non-disjunction of the chromosomes and to chromosome breakage, resulting in the chromosomal abberations that are observed in cell-cultures.

It is more difficult to explain the single gene mutations, which seem to occur at frequencies much higher than the spontaneous mutation frequency in nature. Several mechanisms have been proposed.

• The activation of transposable elements.

Some plant genotypes (eg in maize) have a high spontaneous mutationfrequency because mobile genetic elements are present. The insertion of such an element into a specific gene results in the inactivation of this gene. It is likely that in many plant species silent transposable elements are present, which are mobilized by the tissue culture environment.

- The activation and inactivation of genes by changing the methylation of the DNA.

  It is known that genes which are methylated are not active. When such a methylation is stable and transmitted through meiosis a change in methylation resembles the effect of a mutation.

- Gene amplification and gene depletion.

  Molecular analysis has shown that in some cases the copy number of certain DNA sequences has been increased (amplified). This seems to occur especially with repeated DNA and may have no clear phenotypic effects. The direct cause of this probably rather localised enhancement of replication is not known. Gene depletion is the opposite and may be less common;

- Mitotic crossing over.

  Normally recombination between the two homologous chromosomes, that are present in a diploid cell, takes place only in meiosis. Recombination during mitotic cell divisions is usually thought to be very rare. The effect of mitotic crossing is that if the cell in which this occurred was heterozygous for a particular gene, the cell-line derived from it, will be homozygous for this gene. This is illustrated in Figure 5.3.

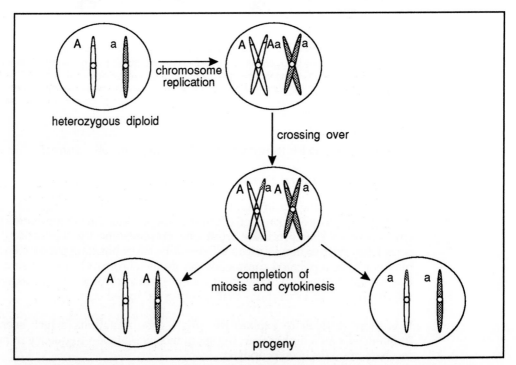

Figure 5.3 Mitotic crossing over in a cell heterozygous for gene A resulting in progeny homozygous for gene A (for simplicity only a single chromosome is illustrated).

There are indications that mitotic crossing over is much more frequent in cultured cells than in somatic cells. This process probably explains why one sometimes finds very high 'mutation frequencies' when the starting material is heterozygous.

It is not assumed that the components of the tissue culture media themselves cause damage to the plant DNA (ie they are not mutagenic).

5.2.5 Factors that modify the amount of variation

From what has been discussed above, it will be clear that it is desirable to control the amount of genetic variation in cell cultures.

The factors known to affect this can be sub-divided into two groups, namely the nature of the starting material and tissue culture factors.

The starting material

Important in this respect are:

- The type of explant.

  The genetically most stable tissues are apical meristems. Multiplication from this tissue in general is the most reliable way of vegetative propagation.

- The plant species.

  Species with relatively low chromosome numbers seem to be more stable with respect to chromosome mutations than, for example, polyploid species.

- Genetic differences within a species.

  It has been observed that significant differences may be present between different genotypes (eg between cultivars) of the same species for chromosome mutations (eg potato and tomato) and single gene mutations (eg between inbred lines of maize).

Tissue culture factors

- Time in culture.

  It has been mentioned before that mutations are found more often the longer the cells are in culture. This is clearly an accumulation effect, which together with the selective advantage or disadvantage of the mutations, determines the ratio of normal to mutant cells present in a culture.

- The type of tissue culture.

  A rather general observation is that tissue cultures that include a callus phase (also all protoplast cultures) show much more genetic variation than for example cultures where no clear dedifferentiation has taken place. Since the type of tissue culture is determined to a large extent by the hormone composition of the media, it is understandable that the type of hormones employed also leads to differences in genetic stability.

There are two alleles which code for product A: one is labelled A, the other a. In a heterozygote carrying both alleles (ie Aa), A is dominant. We can draw this situation as:

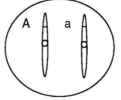

Heterozygous cell carrying A and a

What will be the phenotypic consequence if a transposable genetic element is inserted into the A gene?

Indicate which of the following statements are true.

1) The presence of transposable genetic elements are likely to lead to apparent mutations.

2) Mitotic crossing over is common, whereas crossing over between homologous chromosomes during meiosis is rare.

3) Calli derived from highly differentiated cells are a better source of somaclonal variation than are meristematic cells.

4) Stable changes in gene expression are always the result of mutations (changes to genes).

5) Chromosome number mutations are usually easier to detect than are mutations within the structure of individual chromosomes.

5.3 The selection of mutants from cell and tissue cultures

cell selectable markers

Mutants are not only important for direct application in agriculture or industry but also for fundamental plant research. In addition some types of mutants are useful in particular cell biological experiments such as protoplast-fusion and transformation, where there is a need for selectable markers.

advantage of using cell culture techniques

In cell cultures, a large number of cells in a physiologically uniform environment can be manipulated and screened for rare variants. The use of cell cultures is often cheaper and less labour intensive than handling the same number of plants in a field or greenhouse. Therefore the selection of mutants in cell cultures seems an efficient new tool for plant breeders. However there are a number of problems and complications related to mutant selection in such cultures. These are:

• the selected variants can be due to epigenetic changes and do not express the selected trait after plant regeneration;

• traits that are expressed at the callus stage may not be expressed in whole plants. For example plants regenerated from paraquat resistant calli were not resistant to this herbicide when it was applied to the plants. However calli induced on these regenerated plants were resistant;

- properties which are important for plants are not expressed at the level of cells or undifferentiated tissue such as callus (eg flower colour);

- many mutations are recessive, which implies that these mutations are not detected because the heterozygote cannot be distinguished from the wild type homozygote (ie phenotypically AA - Aa are the same). A solution for this problem is the use of haploid cell cultures in which recessive mutation can be detected directly (A - a). However haploid cell cultures are not available for many plant species and often in these cultures chromosome doubling occurs quite rapidly. But, would it matter if chromosome doubling occurs rapidly? The dihaploid would be homozygous for the recessive mutation and therefore would still show it in the phenotypes.

5.3.1 Should mutagens be used?

limits to the use of mutagens

It is sometimes difficult to avoid genetic variation in tissue culture, in situations where one is interested in stable vegetative propagation. However when searching for specific mutants it is desirable to increase the number of mutations by using treatments that damage the DNA . The mutation frequencies that can be obtained depend largely on the mutagen dose that is applied. However there are limitations to the dose since ultimately cells die when they contain too many mutations. Even at non-lethal doses, regeneration is reduced and the plants with the selected mutation often contain other, unwanted, mutations.

Therefore one has to find a compromise between the desire to produce a high mutation frequency for a particular mutant type and a low number of unwanted mutations. Since mutations also occur 'spontaneously' in cell cultures, it is not surprising that often no mutagen was used to obtain certain types of mutants. In practice a very important consideration is the effectiveness of the selection procedure. When this is very efficient it will be easy to screen large populations of cells. However when the screening procedure is very laborious, a high mutation frequency is often more important than a low frequency of 'background' mutations.

5.3.2 Mutagens

chemicals and irradiation as mutagens

In cell and tissue culture the same type of mutagens that are effective on whole plants have been shown to yield mutations. Examples are the chemicals N-ethyl-N-nitrosourea (NEU), N-methyl-N-nitro-N-nitrosoguanidine (MNNG) and ethylmethanesulphonate (EMS). Effective physical mutagens are X-ray, gamma irradiation and ultraviolet (UV) light. UV light is attractive because it is very easy to apply. A complication is the low penetration of UV light for which cell walls are an effective barrier. Nevertheless UV is an especially favoured mutagen for treatment of protoplasts.

Apart from the type of mutagen and the dose of mutagen that is applied, the factors such as the plant species, the ploidy level of plant or tissue and conditions during the treatment are also important.

A survival curve, obtained by plotting the fraction of surviving cells against mutagen dose, can give an idea about the dose of mutagen that can be applied (see Figure 5.4).

∏ Examine Figure 5.4 and decide a suitable dose of mutagen. Assume that you can treat 10^6 plant cells.

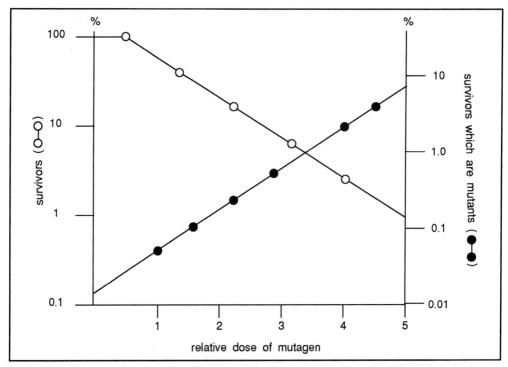

Figure 5.4 Stylised representation of survivors and proportion of mutants after treatment with a mutagen.

The key to working out a suitable dose is to calculate the number of survivors and the number of mutants. We prefer to have a lot of mutants in a fairly small population of survivors. Thus if we did not treat the culture with mutagen we would have 10^6 survivors of which 0.01% would be mutants (ie 100). If we treated the culture with about 3 units of mutagen, we would have about 4% survivors (ie 40 000) of which about 2% would be mutants (ie 800). Which would be the easier to detect 800 mutants in 4×10^4 cells or 100 mutants in 10^6 cells?

If we use very high doses of mutagen, we would have very low survival rates but the proportion of mutants would be high. We would, however, have fewer mutants in total, thereby lowering the probability of producing the actual mutant being sought. The data we used in Figure 5.4 have been stylised, but will enable you to gain some idea of the need to balance mutation rate against survival rate.

An argument for the use of more than one mutagen is, that because of differences in mechanism of actions, the various mutagens can give different types of mutants (ie each has a different mutant spectrum).

5.3.3 The cell culture system to use for mutant selection

It is possible to use different cell and tissue culture systems to induce mutations and to select for mutants. However each system has its advantages and disadvantages.

problems with
cell aggregates

Callus and suspension cultures are attractive because they are easy to induce and to maintain. Since neither types of culture consist of single cells, different physiological properties between cells and between cell aggregates within the same culture exist and may complicate selection. This problem of physiological heterogeneity is more pronounced in a callus growing on solid media and less in suspension cultures. In the latter case the aggregates are often smaller and substances in the medium (eg the selective agents) have a better contact with the cells than within a callus.

Another problem with many callus and suspension cultures is that they have been in culture for a long time and have, therefore, accumulated mutations and epigenetic variants and are difficult to regenerate into plants.

Protoplast cultures have the advantage that they are physiologically and genetically more uniform, especially when the protoplasts are isolated from plants and not from cell-cultures. Even though selection often cannot be applied immediately at the single cell stage, the small microcalli that each derive from single protoplasts are relatively uniform in size and in physiological properties.

feeder layer system A reason that one cannot select immediately after the mutagen treatment has been applied to the single cells is because mutations may need some time for expression. For example: sufficient amounts of a new enzyme may have to be produced before the cell will show a changed phenotype. A technical problem with the selection of a very small number of single cells or small cell aggregates is that plant cells need to be cultivated at a certain cell density to grow well. A common feature of many plant cell suspension cultures is that the cells have to be cultivated at a fairly high density (ie the cells do not grow when incubated at low cell density). We will not go into the reasons for this here except to indicate that the cells supply each other with 'accessory' growth requirements. A way to solve this problem when attempting to grow small numbers of cells is to use feeder-layer systems, where the cells necessary for the support of growth are physically separated (eg by a layer of filter paper or cellophane) from the few selected mutant cells (see Figure 5.5).

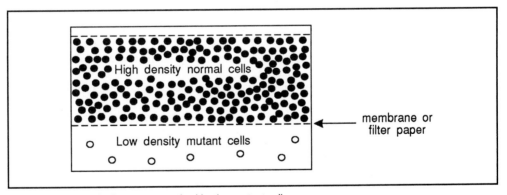

Figure 5.5 The feeder layer system of cultivating mutant cells.

Since it is possible to isolate protoplasts from haploid plants this is the most efficient way to obtain a large number of single haploid cells for mutant selection. Protoplasts are also an efficient way to get cell-clones which can for example be tested for quantitative differences in the specific traits of interest.

5.3.4 The development of mutant screening procedures

When selecting for mutants, in most cases one deals with one of the three following situations:

1)   the mutant grows better under selective conditions than the wild-type. Example: selection for resistance to a toxic compound;

2)   the mutant grows as well as wild type under specific conditions, but grows worse than wild-type under other conditions. Example: the selection for mutants that need a specific compound (eg a vitamin or amino acid) in the medium, which a wild-type makes itself. Such mutants are called auxotrophs;

auxotrophs

3)   the mutant has no selective advantage or disadvantage with respect to growth, but has a distinct phenotype. Example: mutant cells that produce a chemical that can be easily detected (eg the pharmaceutical shikonin which has a red colour).

It is obvious that the first situation is the most attractive one from the point of view of efficiency.

enrichment   Procedures have been developed to change situation 2) into situation 1) by adding compounds to the medium that kill dividing cells but that do not affect non-growing cells. In this way the cell population will become 'enriched' for non-growing cells, which later on can be rescued by adding the chemical that they lack. This procedure has been used to isolate amino acid auxotrophs. Thus, to isolate these, the following procedure can be used. Incubate the mixture of cells (wild type and mutants) in a medium containing no accessory amino acids, but containing a compound which kills growing cells. Under these conditions, the wild type cells will begin to grow and will be killed. The mutants which require particular amino acids will not grow. Thus the proportion of mutants to wild types will be increased.

∏   This can be continued however for only a relatively short period. Why? (Think about what will be released from the killed cells and what effect this may have on the auxotrophic mutants that are present).

In the type of enrichment procedure described above, the wild type cells which are killed will begin to undergo autolysis. This will release amino acids which, of course, will supply the auxotrophic requirements of the mutants. These will begin to grow and, of course, be killed by the same agent that killed the growing wild type cells. After the enrichment the cells are plated out and screened for their amino acid requirements.

In situation three the mutant of interest can often only be detected at the level of the regenerated plant. This is rather similar to the conventional mutagenesis using plants and seeds except that cells and tissue cultures are used to generate the mutants either by applying a mutagen or by using somaclonal variation. The application of mutagens to seeds or shoot tissue is probably easier.

5.3.5 Some examples of mutant selection in plant-cell cultures

With some knowledge about the biochemical pathways of chemical compounds that are made by the plant cells, it is possible to develop screening procedures that allow the selection of mutants which produce more or less of that specific compound.

Example 1: Lysine and methionine are both derived from aspartate (Figure 5.6).

∏   Examine Figure 5.6 and see if you can explain the observation that high concentrations of lysine are toxic to plants.

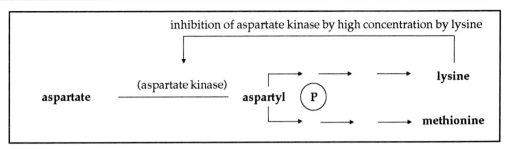

Figure 5.6 Schematic representation of a part of the aspartate pathway of amino acid biosynthesis.

It is explained by the fact that lysine inhibits aspartate kinase (feed-back inhibition), which results in methionine deficiency. A method to overproduce both amino acids is to mutate the gene for aspartate kinase in such a way that the enzyme becomes insensitive to this feedback inhibition.

A way to select for such mutants is to isolate those plants/cells whose growth are not inhibited by high concentrations of exogenously applied lysine.

selection through lack of a specific enzyme

Example 2: Nitrate reductase is an essential enzyme for the metabolism of the nitrogen that is taken up as nitrate by the root system of the plant. It converts nitrate into nitrite. Mutants that lack a functional nitrate reductase enzyme have been isolated in many plants species (and fungi) on the basis of their resistance to chlorate. This is because the same enzyme can also use chlorate as a substrate and converts this compound into the toxin chlorite. In the nitrate reductase deficient genotypes this does not occur.

dominant and recessive mutants

Mutations that lead to a modified enzyme as in the lysine/methionine case described above are often dominant. A 'loss-of-function' mutation (the gene does not make the protein/enzyme any more) as in example 2 however, is almost always recessive. The consequence is that one should use haploid cell cultures (eg protoplasts derived from a haploid plant) to select these mutants.

selection through herbicide resistance

Example 3: Farmers who want to use herbicides for weed-control often cannot use specific herbicides when their crop is growing because the herbicide not only kills the weeds but also the crop plants. Therefore resistance of crop plants to herbicides is attractive from the point of view of agriculture and chemical companies that produce herbicides. Selection for herbicide resistant genotypes at the cell level seems simple because one only has to look for surviving cells in toxic concentrations of the compound. The target of the herbicide is often a protein that is inhibited in its function by the herbicide. It has been possible to modify the target-enzyme in such a way that this does not occur anymore (we will learn of a specific example later in this text).

Many herbicides (eg triazines) affect the photosynthesis system of the plant. When sufficient sugar is provided, photosynthesis does not operate in cell cultures, which explains why even high concentrations of such herbicides are not toxic in these conditions (especially when these grow in the dark). By lowering the sugar concentrations and by growing the cultures in light it is possible to get toxicity to triazine and to select for resistant mutants .

selection through stress tolerance

Example 4: Breeding for stress tolerance is relatively difficult. One reason is that it is not easy to get uniform test conditions in the field. Stress tolerance seems therefore an attractive character to select for in cell-cultures because the stress conditions can be applied here in a very standardised way. It is possible to mimic drought by growing cells on a medium with a high osmotic value and also salts (to select for salt tolerance), and heavy metals can be added very easily to the culture medium. High or low temperatures can be regulated by the temperature switch on the incubator allowing selection for growth at temperature extremes applied to a very large number of 'individuals' (cells).

Surprisingly selection of stress tolerance has been rather unsuccessful, although probably many different selection schemes have been tested in laboratories all over the world. Reasons for this might be that many putative mutants were epigenetic variants and that the relation between properties of cells in culture are different from the behaviour of differentiated cells working together in a tissue of an intact plant. Also the fact that the physiological and biochemical changes that are necessary to get stress tolerance may be complicated and cannot be achieved by one mutation. Many genes may be involved in stress tolerances, and the biochemical pathways are not understood.

selection through a secondary metabolite

Example 5: A potential commercial application of cell cultures is the production of secondary metabolites by large scale cultures growing in fermenters. The first product that was produced commercially in this way by cell-cultures of *Lithospermum erythrorhizon* was shikonin, a pharmaceutical with a red colour. This red colour allows an efficient selection for highly productive cell-lines. In this situation it is not necessary that the high productivity is due to genetic changes because no plants and sexual progeny have to be produced later on. However the relatively unstable nature of epigenetic changes probably explains why it is often difficult to maintain the high productivity of these cell-lines.

SAQ 5.5

Examine the metabolic pathway below carefully.

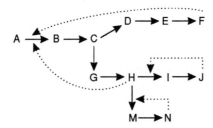

The solid lines represent the pathway, the dotted line indicates feed-back inhibition (eg F inhibits the conversion of A to B). F, J and N are all essential for growth.

1) What compounds would you add to the medium in order to find over-producers of F?

2) If F was brightly coloured, could this be used as a marker for an over producer of J and/or N?

3) Would an over producer of N also be an over producer of F?

Summary and objectives

This chapter has examined genetic variability and the selection of mutants from plants and cultured plant cells.

Now that you have completed this chapter you should be able to:

- describe the differences between epigenetic and genetic variation and explain a strategy for distinguishing between them;

- explain the differences between the terms variants and mutants;

- describe how recessive mutants can be determined by self-fertilisation and the segregation of phenotype characteristics;

- explain how transposable genetic elements, methylation gene amplification and depletion and mitotic crossing over all may contribute to somaclonal variation;

- use knowledge of the advantages and disadvantages of various systems to select suitable in mutant selection procedures;

- use knowledge of metabolic pathways to design selection procedures for the isolation of mutants.

Somatic hybridisation

6.1 Introduction   118

6.2 Protoplast fusion methods   119

6.3 Selection and culture of heterokaryons   126

6.4 Verification of the hybrid or cybrid nature   129

6.5 Asymmetric hybrids   132

6.6 Hybrids and cybrids of interest for plant breeding   133

6.7 Perspectives   134

Summary and objectives   136

Somatic hybridisation

6.1 Introduction

6.1.1 Why somatic hybridisation?

protoplasts
Efficient techniques have been developed for isolation of 'protoplasts' (cells without cell walls) from the tissues of many plant species. Culture of these protoplasts and regeneration of cells to form calli and mature plants has stimulated many scientists to investigate the possibilities of genetic manipulation at the cellular level.

heterokaryons
One approach is somatic hybridisation, involving the fusion of protoplasts from different origins. Regeneration of fusion products, ie heterokaryons (cells with two nuclei containing DNA from two different origins) may result in hybrid plants, which carry genetic information from both parents.

somatic and sexual hybrids
There are clear differences between these somatic hybrids and sexual hybrids obtained via crosses, with regard to the combination of parental characters. In somatic hybrids, nuclei and cell organelles of both fusion partners are combined. In sexual hybrids, cell organelles are only maternally inherited. After protoplast fusion nuclear genomes from both parents are present. This leads to a doubling of the ploidy level (the number of chromosome sets per cell) in cases where cells with the same ploidy level are fused.

SAQ 6.1

What will be the ploidy level of a sexual hybrid obtained after a cross between two diploid plants?

barriers to sexual mating
With the technique of somatic hybridisation, barriers to sexual mating (eg incompatibility, sterility) which exist in plant breeding may be circumvented and transfer of agriculturally important traits, such as disease resistances from wild species, into food crops may be realised. As many of these traits are controlled by more than one gene, often in a complex manner, somatic fusion seems to be the most suitable approach for introduction of these genetic characters. Direct gene transfer through *Agrobacterium* as a transfer vector, or by electroporation (we will discuss this later) of the protoplast plasma membrane is applicable only if the genes to be introduced are known from the molecular point of view and have been cloned into plasmids.

SAQ 6.2

What property is a pre-requisite in hybrid plants obtained from protoplast fusions aimed at introducing important agronomical traits from a wild species into the cultivated species? (Think about what you will want to do with the first hybrid plants that you produce from protoplast fusion).

6.1.2 New genetic recombinants

unique
organelles and
nuclei
combination

cybrids

An alternative to these types of fusions are those in which the nuclei of the donor protoplasts have been eliminated by centrifugation (these protoplasts are called cytoplasts) or have been inactivated by ionizing radiation. In this case, fusion may result in hybrid plants having the original nuclear genes of the recipient and cytoplasms of both fusion partners. These so-called cybrids are of practical interest when agricultural traits coded by cytoplasmic genes need to be transferred.

Somatic fusion involves a number of steps:

- an appropriate procedure for the isolation of protoplasts;

- an efficient membrane fusion resulting in a high proportion of binucleated heterokaryons;

- selection of the heterokaryons after fusion;

- culture of the heterokaryons resulting in a high percentage of dividing cells and regeneration of the putative hybrid colonies to viable plants;

- analysis of the hybrid or cybrid character and genetic constitution of these plants.

A schematic representation of somatic hybridisation is given in Figure 6.1. It must be emphasised that all operations must be carried out aseptically.

6.2 Protoplast fusion methods

6.2.1 Parental protoplasts

Protoplasts used in fusion experiments are mostly isolated from leaf or hypocotyl tissue of *in vitro* shoot cultures and/or from cell suspension cultures. Mesophyll and hypocotyl protoplasts exhibit a better regeneration capacity than protoplasts isolated from cell suspensions. Why is this the case? Remember that due to repeated subculture, cell suspensions have the tendency to adapt to the culture conditions and show an increase in ploidy level or chromosomal abnormalities, resulting in loss of regeneration capacity.

pectinase

cellulase

fusogenic

The enzyme mixture used for protoplast isolation mostly consists of a combination of a pectinase (for separation of the cells) and a cellulase (for cell wall digestion). The choice of the type of enzymes and the duration of the cell wall digestion will determine the viability of the protoplasts and also influence the fusogenic properties.

∏ Write down at least one reason why protoplast preparation influences the efficiency of protoplast fusion.

To get good fusion, the two cytoplasmic membranes need to be brought together. If too little cellulase treatment has been used the membranes will still have cellulose fibrils on their surface thereby preventing fusion. If the protoplasts are kept too long they may become leaky because they are fragile structures and, therefore, lose viability.

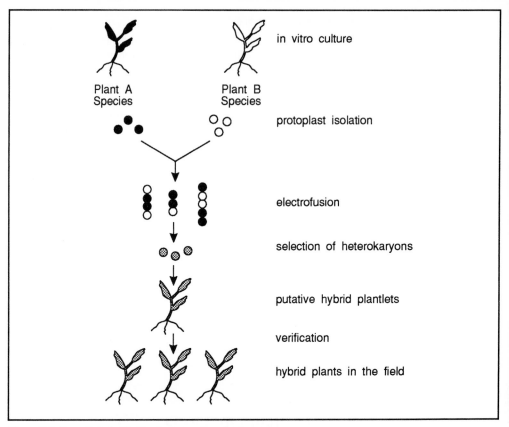

Figure 6.1 Somatic hybridisation between two plant species.

visualisation of heterofusions

It is advantageous to follow the events during and after fusion and to evaluate the percentage of fused cells. Thus, heterofusions should be recognisable visually. In a fusion combination of mesophyll protoplasts containing green chloroplasts and cell suspension (colourless) protoplasts, showing cytoplasmic strands, heterofusions can easily be identified under the microscope. Colourless protoplasts can also be isolated from bleached plantlets. These can be obtained by growing plants on a medium with a herbicide. Use of fluorescent dyes may facilitate the identification of heterofusions.

6.2.2 Chemically induced fusion

agglutination with PEG

As protoplast plasma membranes carry a net negative charge on the surface, protoplasts tend to repel each other. For fusion a close contact between the membranes is required. Therefore, chemical fusion methods aim at modifying the surface charge by pH change of the medium, polycations (eg calcium) or by dehydrating effects. The widely applied polyethylene glycol (PEG) fusion method will result in agglutination of the protoplasts by binding with phospholipids within the plasma membrane. The agglutination is dependent on the temperature of the medium, molecular weight and concentration of PEG and the duration of the PEG treatment. Subsequent dilution with a $CaCl_2$ containing medium at high pH stimulates fusion of the membranes. After careful washing, the protoplasts and fusion products can be cultured. This sequence is stylised in Figure 6.2.

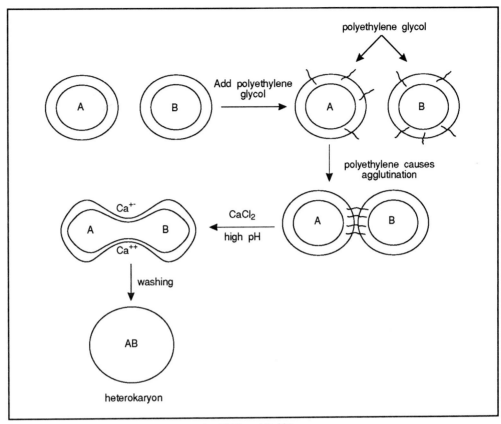

Figure 6.2 Cell agglutination and fusion using PEG and Ca^{++} ions.

ΙΙ Why should the agglutination of the cells be dependent upon the molecular weight and the concentration of PEG? (Examine Figure 6.2 and think what would be the consequences if the PEG molecules were very short and if the PEG molecules were present in high concentration).

Short chained PEG would, of course get absorbed into one cytoplasmic membrane but not protrude into the environment and could not, therefore bind with other cells. If a very high concentration of PEG was used, each cytoplasmic membrane would be filled by PEG molecules and there would be no room to absorb the protruding ends of PEG molecules tied into other cytoplasmic membranes. Graphically we might expect the amount of agglutination to be influenced by PEG concentration as illustrated in Figure 6.3. Theoretically we would expect an optimum value which would be dependent upon the protoplast concentration used. (Why?). Note that concentrations of PEG above 40-50% are detrimental to protoplast survival.

cytotoxicity of
PEG

chromosome
elimination

The fusion procedure should be optimised for the specific cell lines used. Aggregation of fused protoplasts should be avoided but this will be difficult. The cytotoxic effect of the PEG treatment for the cell lines used should also be evaluated. Fusion frequencies are generally in the range of 0.1 to 2 per cent, while protoplasts with small vacuoles seem to fuse easier than large-vacuolar protoplasts. After fusion of the plasma membranes,

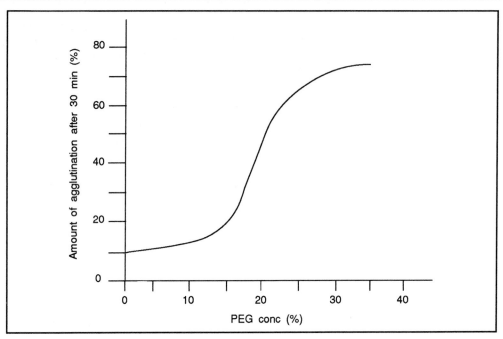

Figure 6.3 Graphical representation of amount of agglutination against PEG concentration (stylised).

nuclear fusion will occur in most cases. Chromosome elimination is often observed. There are indications that specific chromosomes are preferentially eliminated.

6.2.3 Electrically stimulated fusion

electrofusion

Nowadays an alternative fusion method is often used. This is electrofusion. It is a rapid and simple technique which leads to high fusion frequencies up to 10 per cent. Protoplasts become polarised in an alternating current (AC) field of about 1 MHz and a field strength of 50 to 100 V/cm. They will move to positions of higher field strength, a

dielectrophoresis

phenomenon called dielectrophoresis. In a fusion chamber with thin wire electrodes protoplasts will move towards the wire and form 'pearl' chains. In the case of a chamber equipped with parallel plate electrodes a homogeneous field is expected, but the mere presence of the protoplasts creates local positions of higher field strength. This results in a movement of the protoplasts towards each other ('mutual dielectrophoresis'). Thus, pearl chains appear in the total volume of the chamber parallel to the field lines (Figure 6.4).

Dielectrophoresis is related to the dielectric constant (defined by ε in the equation: $F = QQ'/\varepsilon r^2$ where F is the force of attraction between two charges Q and Q' separated by a distance of r in a uniform medium) of the protoplasts relative to the medium. It is dependent on the frequency of the AC field, the ionic strength of the medium and the diameter of the protoplasts. At high AC fields (0.1-10 MHz) protoplasts seem to be polarised due to polarisation of the plasma membrane. Dielectrophoresis is facilitated in a low conductivity medium because dielectric constants of protoplasts and medium will differ more and protoplasts will produce higher field distortions. Large protoplasts are more susceptible to dielectrophoresis than small protoplasts.

Membrane fusion is induced by a short (10-50µs) direct current (DC) pulse of 1.5-2.5 kV/cm causing 'reversible membrane breakdown' (Figure 6.4). The critical membrane voltage needed for membrane breakdown is first reached at the poles, the area of cell to cell contact. Longer pulses or higher DC voltages will cause membrane breakdown also outside the cell contact area.

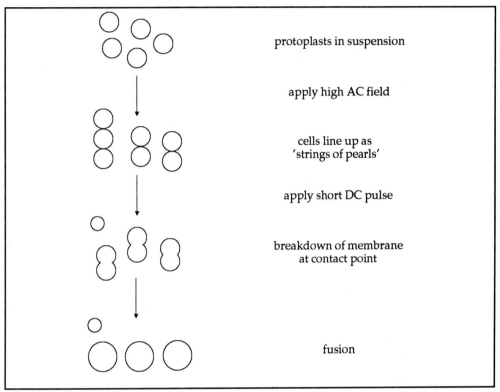

Figure 6.4 Electrofusion - protoplasts are lined up in a short string of pearl arrangement in an AC field. A short DC pulse of moderate voltage causes membrane breakdown and fusion of the protoplasts.

SAQ 6.3	Why should longer pulses or very high DC voltage be avoided in carrying out electrofusions?

6.2.4 The practice of electrofusion

Electrofusion is carried out in fusion chambers with volumes up to 0.7 ml and 10^5 to 10^6 protoplast/ml. The fusion process can easily be followed under the microscope. Fusion parameters can be adjusted depending on the actual fusion conditions (quality of the protoplasts, conductivity of the medium). The fusion frequencies obtained with electrofusion are reproducible and are in the range of 1 to 10%.

Physical parameters can be chosen in such a way that mainly binucleate heterofusions are produced (see Figure 6.5 for example).

Π Which is the best DC pulse voltage to use? For this use Figure 6.5 and think about what we are trying to produce (binucleate or multinucleate cells).

According to the data presented in Figure 6.5, we would have selected a DC pulse of about 500 volts/cm. This gives a high percentage of fusions producing binucleate cells, but a low percentage of multinucleate products.

Cell lysis should be avoided as much as possible by inducing only membrane breakdown at the area of cell to cell contact.

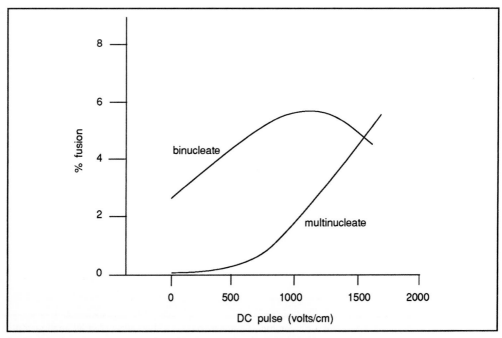

Figure 6.5 Proportions of bi- and multinucleate cells after electrofusion.

Electrofusion should be carried out in a non-conductive medium (eg mannitol) to prevent movement of the protoplasts in convection currents caused by the heat effects in the medium (Joule heating) resulting from the passage of electricity. The reduction in convection currents will facilitate protoplast alignment.

Π Write down at least one reason why electrofusion is not carried out with protoplasts in distilled water.

A correct osmolarity of the non-conductive medium is needed in order to keep the protoplasts in a viable and fusogenic state. If the medium is not isotonic with the cell sap they will either shrink (hypertonic) or swell and possible burst (hypotonic).

Problems may arise when cell membranes are not in close contact with each other due to incomplete enzyme digestion of the cell walls. Also fusion combination of vacuolar and non-vacuolar protoplasts each having a different specific weight, may reduce the number of heterokaryons formed.

Π Explain why protoplast fusion is hampered under these conditions.

You probably realised that one set of cells will tend to settle more quickly than the other resulting in parts of the suspension being rich in only one cell type.

Indeed, pearl chains consisting of mainly one type of protoplasts will occur. Protoplasts having different densities will tend to be separated from each other.

In general, mesophyll protoplasts have more fragile cell membranes than protoplasts from cell suspensions. Addition of calcium ions during fusion may have a positive effect on membrane stability and may result in higher fusion frequencies.

SAQ 6.4	List as many the advantages and disadvantages of the two fusion (chemical and electrofusion) methods as you can.

PEG fusion   Advantages

                                  Disadvantages

Electrofusion   Advantages

                                  Disadvantages

6.2.5 Microfusion

The fusion methods discussed so far involve mass fusion of protoplasts. It should be noted, however, that the parental protoplast populations are heterogenous with respect to cell cycle, physiology, morphology and regenerating capacity. A recently developed system for individual selection, culture and regeneration of plant cells offers possibilities for fusion of pre-selected protoplasts. Also with this microfusion technique, fusion products have been obtained from which whole plants were regenerated.

pre-selected protoplasts

The selected protoplasts are transferred into a microdroplet of low ionic strength fusion medium, deposited on a microscope cover slip (Figure 6.6). Electrofusion is performed with micro-electrodes which are brought near to the protoplasts. Several protoplast containing microdroplets are placed on the coverslip in a fusion experiment. They are covered with a layer of mineral oil to prevent evaporation. After fusion, the fusion products are transferred to a microculture chamber for individual culture. This microfusion technique also allows fusion of different cell types, such as between protoplasts and cytoplasts (protoplasts without a nucleus).

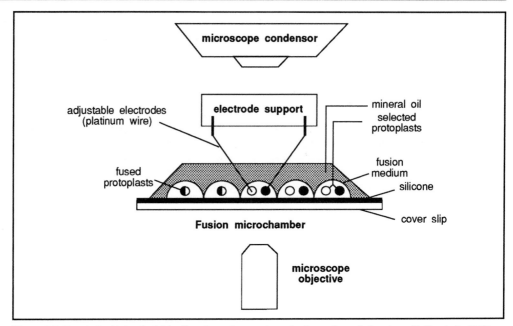

Figure 6.6 Set-up for fusing individually selected protoplasts (redrawn from Schweiger, H. G. *et al* , 1987, Theoretical and Applied Genetics 73 769-783).

6.3 Selection and culture of heterokaryons

6.3.1 Visible markers

Several methods have been developed for the selection of fusion products. The obvious one involves the inability of the parental protoplasts to divide and proliferate on the culture medium chosen, while heterokaryons can undergo cell division and further growth. This is however, by no means a general approach towards selection. In the *hybrid vigour* fusion of protoplasts, that are of related origin, hybrid vigour (faster, more vigorous growth of hybrids) may be helpful in the selection process. Somatic hybrids have been obtained in this way.

Microfusion and microculture do not need any selection criteria since single fused products are involved. After mass fusion of protoplasts active selection of heterofusions *morphological markers* is required. Morphological markers are therefore needed (eg the presence of chloroplasts in mesophyll protoplasts and cytoplasmic strands in protoplasts from cell suspensions).

fluorochromes Fluorochromes can be used to distinguish heterofusions from parental protoplasts. For example, mesophyll protoplasts show red chlorophyll fluorescence, and fluorescein diacetate (FDA) stained protoplasts appear yellow green. Heterokaryons from fusions of these types of protoplasts can be identified within 3 days after fusion by the presence *micro-* of both types of fluorescence. The heterokaryons can be collected with a micropippette *manipulation* and micromanipulator. In skilled hands 50 to 80 per h can be fished.

<table>
<tr><td>

SAQ 6.5

</td><td>

Why is it of advantage to collect fusion products after some days instead of directly after fusion has taken place? (Think about what will happen to the protoplasts after they have fused).

</td></tr>
</table>

6.3.2 Cell sorting

fluorescence
activated cell
sorter

Heterokaryons can also be sorted automatically with a fluorescence activated cell sorter. For this purpose parental protoplasts each having a different fluorescent label (eg red and yellow fluorescence) are fused and the sample is fed into the flow sorter (Figure 6.7).

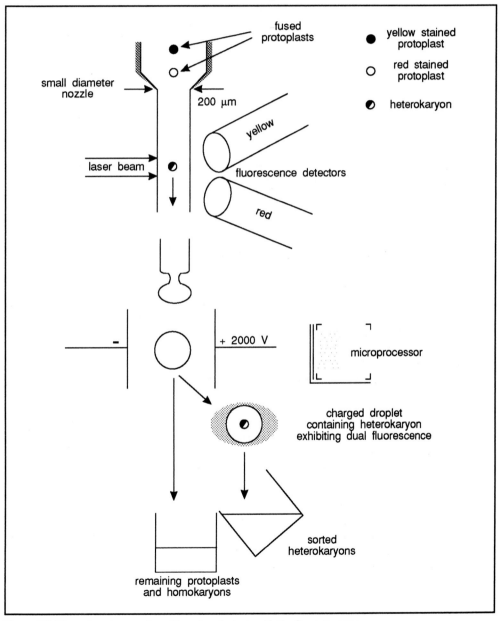

Figure 6.7 The cell sorter - sorting of fused protoplasts with the flowcytometer.

The protoplasts pass through a nozzle with small diameter (200 μm) orifice, which narrows the sample stream and centres the protoplasts at the stream axis. The stream is intercepted by a laser beam which excites the fluorescent label of the protoplasts when they pass this measuring point. The fluorescence signals are fed into two photomultipliers, which will register the red and yellow signals, respectively. Simultaneous registration of red and yellow signals activates a microprocessor resulting in formation of a number of electrically charged small droplets at the end of the sample stream. One of these droplets will contain the protoplast exhibiting the 'dual fluorescence', probably a heterokaryon. These charged droplets pass through an electric field, are deviated and will be collected in a separate tube.

It should be noted that only a fraction (10-40%) of the sorted heterofusions can be recovered intact. Many of these will burst during passage through the nozzle orifice. Only heterokaryons having a not too large diameter and a flexible plasma membrane (cell wall synthesis not yet started) can be recovered. Nevertheless, many intact heterokaryons can be collected in a short time (250-1500 per hour). In practice, the proportion of recognisable heterofusions in the sorted sample is about 80%.

6.3.3 Biochemical markers

selection by biochemical markers

Hybrid selection can sometimes by based on the presence of biochemical markers. Mutant cell lines have been isolated which are auxotrophic for certain chemical compounds in the culture medium or which are resistant to certain drugs. Fusion of protoplasts from mutant cell lines both having a different recessive selectable trait may lead to genetic complementation of the genome by the fusion partner and offers the possibility for selection of the hybrids. Hybrid selection after fusion of cells with different deficiencies in nitrate metabolism is a good example. Only hybrids will be able to grow on medium with nitrate as the sole nitrogen source.

SAQ 6.6	Incorporation of an auxotrophic trait and a drug resistant trait in the same cell line allows hybrid selection after fusion of these mutant cells with cells from any non-mutant cell line. What should the culture medium contain for selection of these hybrids?

Hybrid selection based on the use of mutants has the advantage that often a simple selection scheme can be followed. However, isolation and characterisation of these mutants is time consuming. Mutants are often not desirable in practical fusion applications because they produce undesirable characteristics in the product (eg plants that require special growth factors).

6.3.4 Culture of heterokaryons

Culture and regeneration of heterokaryons is an essential step in obtaining hybrid or cybrid plants. Fortunately, regeneration procedures have been developed now for many plant species. Plant regeneration from heterokaryons can be expected when at least one of the parents exhibits a good regeneration capacity and thus plant regeneration from heterokaryons will mainly depend on the origin of the plant material used in fusions.

culture at low cell density

Cultures of heterokaryons collected via a micromanipulator or with a cell sorter need special measures. Protoplasts are commonly cultured at a density of 1 to 5×10^4 per ml in petri dishes of 1.5 ml culture medium. In order to culture low numbers special techniques have to be applied. Use is made of feeder cells, agarose beads, culture in small droplets etc. These same systems are in principle suited for culture of low numbers of heterokaryons.

6.4 Verification of the hybrid or cybrid nature

6.4.1 Cytological analysis, isoenzymes

type of
evidence
collected to
prove a hybrid
has been
produced

Although fusion products representing fused parental protoplasts may have been collected, a hybrid or cybrid character of calli and plants regenerated from these heterofusions is by no means guaranteed. Indeed, it is not certain that after protoplast membrane fusion, nuclear hybridisation will also occur. It is also possible that parental protoplasts may erroneously be collected instead of heterofusions. A presumed hybrid vigour, as well as a correct chromosome number (ie the sum of the parental chromosomes), or the morphology of the regenerated plants can be no more than an indication of the possibility of a hybrid nature. Identification of both types of parental chromosomes, using Giemsa banding of the chromosomes of a putative hybrid plant will clearly prove hybridity (NB Giemsa is a stain which will colour chromosomes to give a distinct banding pattern). Furthermore, information on loss of specific chromosomes is obtained with this technique.

Giemsa
banding

isoenzymes

A widely used method is the analysis of 'isoenzymes'. A difference in isoenzyme patterns between parental material may allow verification of hybrids. It should be realised that chromosome loss will have consequences for detection of gene products normally encoded for by genes located on these chromosomes. Analysis of several isoenzymes should be carried out for verification of the hybrid nature.

6.4.2 RFLP analysis

RFLP

Recombinant DNA techniques also offer possibilities for verification of hybridity. The analysis is based on 'restriction-fragment-length-polymorphism' (RFLP). We will examine this technique more fully in the next chapter. For now we provide a summary of the basic ideas. DNA of two different parents may yield DNA fragments which differ in length after digestion with restriction enzymes. These can be separated by electrophoresis. Using a species-specific DNA-probe (eg a radioactive labelled DNA fragment), a species-specific pattern (fingerprint) may be obtained after binding (hybridisation) to the DNA fragments from each parent. In this way hybridity can be checked.

fingerprint

hybridisation

Use of probes, which hybridise to DNA well spread over the genome may give an estimation of how much of the parental genome is present in the hybrid. For a number of agriculturally important crops (eg maize, potato) RFLP 'linkage maps' are being constructed. They provide a direct method for selection of desirable genes via their linkage to RFLP markers (see Chapter 7).

6.4.3 Fate of cell organelles

In sexual hybridisations, the cell organelles such as chloroplasts and mitochondria are uniparental (maternally) inherited. After cell fusion, however, chloroplasts and mitochondria from both parents are present and this situation will finally, after a number of mitotic divisions, lead to different nucleo-organelle combinations. It seems that chloroplasts and mitochondria segregate independently but, in the case of mitochondria, recombination may occur. Recombination of chloroplasts is very rare and it has been reported only once.

SAQ 6.7

The electrophoretic isoenzyme pattern for enzymes A, B, C for two cell lines are drawn below.

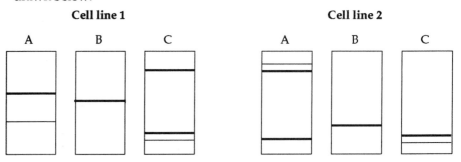

Two suspected hybrids (X and Y) were isolated from the fusion of protoplasts derived from these cell lines. The electrophoretic isoenzyme patterns for enzymes A, B, C for these two hybrids are drawn below.

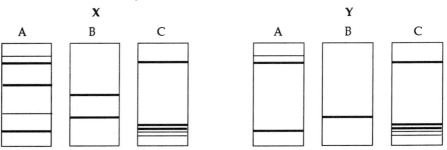

1) Are X and Y hybrids of cell lines 1 and 2?

2) Can you conclude anything about the ploidy state of X and Y?

3) How could you verify the conclusions you came to in 2)?

The direction of segregation is influenced by many factors such as the number of organelles, their replication potential, incompatibility and the time period after fusion.

chloroplast markers

mitochondrial marker

The large subunit (LSU) of ribulose bisphosphate carboxylase, a herbicide resistance (atrazine), antibiotics resistance (tentoxin, streptomycin) and the albino phenotype are excellent chloroplast markers. The agriculturally important trait of cytoplasmic male sterility (cms) is located on mitochondria.

fate of chloroplast

With chloroplasts, a rapid and random segregation of one type of the parental genomes usually occurs. Thus two populations may arise, one will carry chloroplast DNA from one parent cell, the other will carry the chloroplast DNA from the other parent cell. Unidirectional segregation may occur as in the case of nucleo-cytoplasmic incompatibility. In this case one of the donor chloroplasts DNA is incompatible with the nuclear DNA and this combination dies. In addition, the occurrence of mixed chloroplast genomes after fusion has been reported. These are illustrated in Figure 6.8.

Thus two populations may be present for some time until segregation of one population. One will carry chloroplast DNA from one 'parent' cell, the other will carry the chloroplast DNA from the other 'parent' cell.

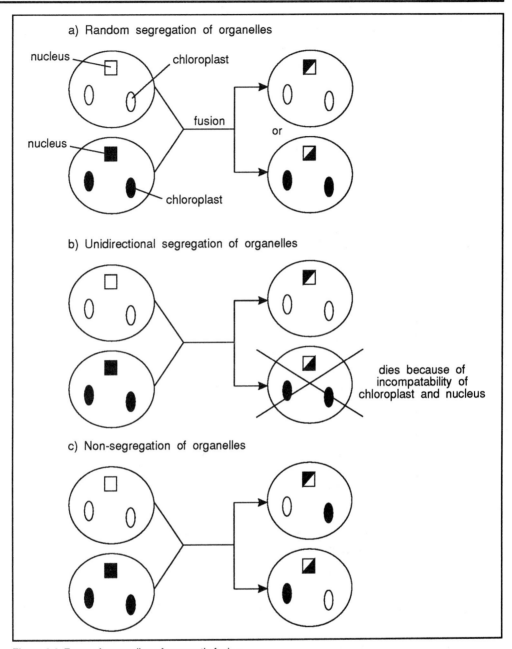

Figure 6.8 Fates of organelles afer somatic fusion.

recombination of mitochondria

The situation for the mitochondria is not clear. The unique DNA restriction profiles (fingerprints) found after fusion have led to the assumption that recombination of the parental mitochondrial genomes occurs. However, in a number of fusion experiments sorting out to one type of parental genome has also been reported. It should be realised that within the mitochondrial genome, which consists of circular DNA molecules, recombination can easily occur, leading to new DNA arrangements. The fate of the mitochondrial genome will certainly be dependent on the mitochondrial DNA stability.

This may be species dependent. Mitochondrial recombination events have also been observed after prolonged *in vitro* culture.

6.5 Asymmetric hybrids

6.5.1 Spontaneous chromosome elimination

As already mentioned, spontaneous chromosome elimination may occur after cell fusion, even when the fusion partners are closely related. The mechanism of chromosome elimination is still obscure. It has been observed that loss of one or a few chromosomes may still produce fertile hybrid plants.

chromosome elimination more common with distantly related fusion partners

Unilateral chromosome elimination will occur when the parents are more distantly related. This results in asymmetric hybrid plants. In several cases regeneration could only be realised after elimination of several or all chromosomes of one parent. The direction of chromosome elimination cannot be predicted beforehand. For example, in somatic hybrid calli of *Solanum tuberosum* and *Nicotiana plubaginifolia* both types of hybrids were present, ie one with an incomplete set of potato chromosomes and an excess of *Nicotiana* chromosomes and the other type with a few *Nicotiana* but an excess of potato chromosomes.

6.5.2 Induced chromosome elimination

chromosome loss and irradiation dose

Chromosome elimination of one of the fusion partners can be induced by X- or γ radiation. The degree of chromosome loss is not directly related to the applied radiation dose. Fusion with irradiated donor protoplasts will lead to fusion products with incomplete donor chromosomes. Loss of donor chromosomes and chromosome fragments carrying a centromere occurs. With *in situ* DNA:DNA hybridisation using species-specific DNA probes, it has been shown that large chromosome fragments can be integrated into the intact chromosomes of the recipient partner.

Several asymmetric hybrid plants have been produced which carry only a few chromosomes or a specific trait from the donor parent (eg *Nicotiana plumbaginifolia* with some *Atropa belladonna* chromosomes, *Nicotiana tabacum* with kanamycin resistance from *N. plumbaginifolia*). Transfer of genetic information from the donor to the progeny will require recombination between the donor and recipient chromosomes leading to introgression of genetic donor material into the recipient genome.

6.5.3 Transfer of cytoplasmic traits

The donor-recipient system with the donor protoplasts being inactivated by X- or γ radiation, as described in 6.5.2, was originally developed for transfer of cytoplasmic traits from the donor. In general, the recipient protoplasts are treated with iodoacetate before fusion to inactivate the cytoplasm. After fusion, the parental protoplasts will not be able to survive and only fusion products will divide and proliferate due to metabolic complementation. With this system, substitution of cytoplasm can be realised and successful transfer of cytoplasmic male sterility - a trait conferred by the mitochondria - has been obtained in a number of crops (*Brassica, Daucus*).

| SAQ 6.8 | Why do the parental donor protoplasts not survive in the procedure described above? |

6.5.4 Microprotoplasts

separation of chromosomes from nuclei

Although little data are available from fusion experiments with microprotoplasts, it is still worthwhile to discuss these. Microprotoplasts are obtained from cells in which 'micronuclei' were formed. During the normal cell cylce, chromosmes condense and become attached to the spindle. For the formation of micronuclei, the attachment of chromosomes to the spindle is disrupted by toxins. After treatment with 'spindle toxins', such as amiprophos-methyl (APM) and oryzalin, chromosomes become scattered throughout the cell. They form groups consisting of two or more chromosomes. The chromosomes de-condense and form micronuclei surrounded by a nuclear envelope. Ultracentrifugation of micronucelated cells results in the isolation of microprotoplasts containing only a few chromosomes. The foreseen use of microprotoplasts in fusion experiments is that it would be an ideal way of partial genome transfer.

SAQ 6.9

What will be the advantage of this transfer system as compared to the donor-recipient system mentioned in 6.5.2?

6.6 Hybrids and cybrids of interest for plant breeding

6.6.1 *Cruciferae*

movement of research from model systems to agriculturally important ones

Recently a shift can be noticed from using model-plants such as *Nicotiana* and *Petunia* for cell fusions towards the use of agriculturally important plant species. This is mainly because of improvement in plant regeneration procedures of many crop plant species. Fusions have been performed within the *Cruciferae* and *Solanaceae*.

With respect to the cruciferous crops, the successful re-synthesis of rapeseed (*Brassica napus*) by fusing *B. campestris* and *B. oleracea* was one of the first results reported. It is interesting to note that in the hybrids obtained from sexually crossable parents, only 30% had the expected chromosome number of 38. Most of these hybrids were fertile. Also fusions between other *Brassica* species have been performed (eg *B. napus* with *B. nigra*) resulting in hybrids several of which had the expected chromosome number and were proved to be fertile.

Fusion of *Brassica* species with *Sinapsis alba* and *Eruca sativa* (the species from two related genera which cannot be sexually crossed with *Brassica*) also yielded fertile hybrids. However, a great variation occurred in the chromosome number in these hybrids.

Many novel nucleo-cytoplasmic combinations have been produced within *Brassica napus*, aiming at the introduction of important traits such as cytoplasmic male sterility and resistance to the herbicide, atrazine, into certain cytoplasms of *B. napus*. As male sterility can also be caused by aneuploidy, further analysis using DNA probes is required to verify if the desired hybrids were indeed obtained. Useful new nucleo-cytoplasmic combinations have been incorporated into breeding programmes.

6.6.2 *Solanaceae*

Many somatic fusions have been carried out between potato (*Solanum tuberosum*) and wild relatives in order to transfer disease resistances into existing potato breeding lines. Somatic hybridisation offers good perspectives especially for wild *Solanum* species which cannot be crossed sexually with potato.

disease
resistances of
donor and
recipient
strains passed
on to hybrids

additional
desirable
features
observed

Using somatic hybridisation a combination of a tetraploid cultivar of *S. tuberosum* (which has resistance to the fungal disease *Phytophthora infestans*) with the diploid *S. brevidens* (a wild potato being resistant to potato leaf roll virus, PLRV, and which does not form tubers) fertile hexaploid hybrid plants were obtained exhibiting both resistances. Unexpectedly, tubers of the somatic hybrid proved to be resistant also to soft rot, a trait not present in the *S. tuberosum* parent. First and second crosses were made with two potato cultivars and it was observed that the first mentioned two resistances could be passed on to their sexual progeny. It is not yet known with certainty that recombination between the two genomes has also occurred.

Superior clones of *S. tuberosum* have also been used for fusions in order to obtain symmetric tetraploid hybrids which should be directly valuable in potato breeding. Some good yielding hybrids with the expected chromosome number have indeed been obtained, but many hybrid regenerants exhibit chromosome loss or increase in ploidy level.

6.7 Perspectives

From the examples given in 6.6.1 and 6.6.2 it is clear that somatic hybridisation can offer a good alternative to breeders where reproductive (sexual) barriers exist. Production of symmetric hybrids with a chromosome number being the sum of that of the parents will be more successful if the parents are more closely related genomically. When both parents are fertile, fertile hybrids may also be expected. These can then be introduced into breeding programmes. In Table 6.1, a list of fertile hybrid plants obtained after somatic fusion is given.

Combination of more distant related species, eg species from different genera or family results, in general, in spontaneous elimination of chromosomes, preferentially from one of the parents. In these cases with unbalanced chromosome numbers, when plant regeneration can be obtained, regenerants exhibit sterility. Elimination of chromosomes from one of the fusion partners can be stimulated by X- or γ radiation. With this procedure, selectable traits from the donor can be transferred, as shown in fusions of irradiated *Daucus carota* with *Nicotiana tabacum* and of irradiated *S. pinnatesectum* with *S. tuberosum*. Experience with the production of asymmetric hybrids is still very limited, but there is a growing interest in partial genome transfer by asymmetric hybridisations and also in cybridisation.

Solanum brevidens (+) S. tuberosum (diploid)	Fertile pollen and eggs at or near the expected tetraploid level	Austin et al, 1985
	Female fertile 3 of 15 hybrids had the expected 48 chromosomes	Fish et al, 1988
Solanum brevidens (+) S. tuberosum (tetraploid)	Female fertile at or near the expected hexaploid level	Helgeson et al, 1986
Nicotiana glauca (+) N. langdorfii	Fertile, viable seeds after self pollination 60 to 66 chromosomes instead of 42	Morikawa et al, 1987
N. tabacum Daucus carota (+) (gamma-irradiated)	Only female fertile variable chromosome number	Dudits et al, 1987
N. tabacum N. plumbaginifolia (+) (gamma-irradiated)	Only female fertile most of N. tabacum cells had 1 extra chromosome	Bates et al, 1987
N. plumbaginifolia Atropa belladonna (+) (gamma-irradiated)	Partly fertile 6 to 29 Atropa chromosomes present	Gleba et al, 1988
Brassica campestirs (+) B. oleracea	Mostly fertile 30% had the expected 38 chromosomes	Sundberg et al, 1987
Solanum phureja (+) S. tuberosum	Mostly fertile at or near the expected 48 chromosomes	Puite, Mattheij, 1989
Brassica napus (+) B. nigra	Mostly fertile 20 of 30 hybrids had the expected 54 chromosomes	Sjodin, Glimelius, 1989
Brassica napus (+) B. juncea and B. carinata (both x-irradiated)	Seed set of selfed hybrids about 10% of B. napus had 38 to 66 chromosomes	Sjodin, Glimelius, 1989

Table 6.1 Some fertile hybrid plants obtained after somatic hybridisation. Note that we have given brief details of those who have carried out successful somatic hybridisation together with the date when this success was recorded. It will give you some idea of how quickly this technique is being developed.

Summary and objectives

This chapter has described the techniques that can be used to achieve fusion between two plant protoplasts and describes the likely outcome of such fusions. It also examines the prospect of these techniques of producing new and desirable plant cultivars.

Now that you have completed this chapter you should be able to:

*   explain the advantage and limitations of using somatic hybridisations;

*   describe the phases in preparing protoplasts and using them for somatic hybridisation;

*   list the many factors which influence the efficacy of polyethylene glycol and pulsed electric fields for stimulating protoplast fusion;

*   list the advantages and disadvantages of chemical and electrofusion techniques;

*   use electrophoretic isoenzyme patterns to determine the hybrid status of protoplast fusion products;

*   explain how the use of auxotrophic and drug resistant mutants can be used to select hybrid products derived from protoplast fusion.

In the next chapter, we will explore the important topic of gene mapping.

Gene mapping and gene isolation

7.1 Introduction   138

7.2 Linkage and crossing over   138

7.3 Restriction Fragment Length Polymorphism (RFLP) analysis   152

7.4 Gene isolation   158

7.5 Gene cloning by the polymerase chain reaction (PCR)   168

7.6 Gene isolation with a heterologous probe   170

7.7 cDNA or genomic DNA?   170

7.8 Chromosome walking   172

7.9 Transposon tagging   174

7.10 Differential screening of cDNA libraries to obtain process specific
     genes   175

7.11 Linking the gene fragment to the trait: definite proof   177

Summary and objectives   182

Gene mapping and gene isolation

7.1 Introduction

∏   List as many ways as you can in which gene mapping may help the plant breeder.

linkage or
segregation of
characters

development
of screening
procedures

detection of
silent genes

gene isolation

There are in fact many ways. For example, you could have noted that knowing whether or not genes are closely linked together will help to decide what strategy to adopt to produce genetic combinations. Knowing whether or not particular genes are close together will give an indication, whether or not particular combinations of traits will be stable or likely to separate in sexually produced progeny. Alternatively, mapping genes to particular chromosomes may enable the breeder to develop screening procedures for identifying progeny carrying (not carrying) particular genes or groups of genes. You may have also thought of using knowledge of gene linkages to help to detect or monitor silent (recessive) genes or genes whose products are difficult to detect. For example, linkage of a dominant drug resistance gene to a recessive gene, would enable drug resistance to, at least in part, monitor the occurrence of the recessive gene. Gene mapping also has relevance to gene isolation, a prelude to using recombinant DNA technology for plant breeding. Whatever reasons you listed, you will have convinced yourself that gene mapping is an important aspect of plant breeding.

This chapter deals with both traditional and contemporary techniques of analysing gene linkage. Biotechnology both contributes to, and benefits from, gene mapping. The development of molecular biological techniques has enabled gene linkages to be analysed using DNA fragments (Restriction Fragment Length Polymorphism analysis or RFLP). We will learn how this technique is applied in a later section. This technique is however an important new contribution and has done much to facilitate gene mapping. We complete the chapter by exploring the technique available for isolating single or small groups of genes.

We begin by reviewing the processes of conventional gene mapping. We have assumed that you are familiar with meoisis and mitosis and will not discuss them in detail here (see the BIOTOL text 'Infrastructure and activities of cells'). Nevertheless this chapter is quite long so we advise you to study gene mapping first and then to examine gene isolation after you have had a break.

7.2 Linkage and crossing over

The phenotype (appearance of an organism) is determined by the genotype. It was established in the first twenty years of this century that genes are located within the chromosomes in a linear order. Genes are therefore transmitted to the next generation in groups. Essentially every chromosome represents such a linkage group.

In the diploid organism, two mechanisms are available for recombining the genetic material. Paternal and maternal chromosomes will be randomly distributed during the formation of the haploid gametes, leading to the presence of one complete set of chromosomes in both gametes. We illustrate this in a simple way (Figure 7.1) using two cell lines which contain a haploid number (N) of chromosomes of one. If you examine this figure, you will see that there are a variety of combinations of chromosomes that can be generated in a sexually produced diploid progeny.

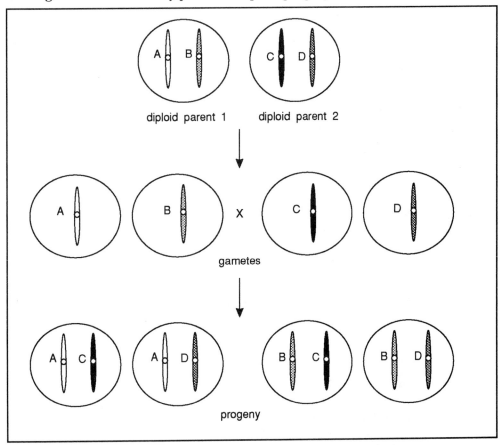

Figure 7.1   Illustration of randomly distributed chromosomes during sexual reproduction.

The number of combination of chromosomes that can be produced in a gamete in a diploid is in fact $= 2^N$ where N is the haploid number of chromosome. Think about this for a moment. In a diploid each homologous chromosome is present in two copies of each - one from each parent. Thus, there are 2 choices for each chromosome that are incorporated into gametes. Therefore it follows that if, for example, the haploid number of chromosomes is 4, then there are 2 x 2 x 2 x 2 combination of chromosomes possible in the gametes.

∏   How many combinations of chromosomes are there in the gametes of a diploid which contains 10 chromosomes?

The answer is 32. Remember that of the diploid contains 10 chromosomes, the haploid number is 5. Therefore the number of combinations of chromosomes is 32 (ie 2 x 2 x 2 x 2 x 2).

∏   In such a diploid, how many combinations are possible in the zygotes?

random sorting of chromosome leads to a large number of chromosome combinations

Your instinct may have been to say 1024 (ie 32 x 32). In principle this is correct. In practice however if any of the chromosomes found in the diploid parents are identical the answer would be less. The calculation of possible combinations of chromosomes, would be further complicated by the presence of sex chromosomes. It is not our intention to cover these here. The message we have intended to convey is that the random sorting of chromosomes in gamete formation produces a wide range of chromosome combination. In principle by analysing the occurrence of phenotype characteristics in the progeny we can determine whether or not these characteristics are carried on the same chromosomes. We will carry out such analysis in the next section.

crossing over

The second mechanism for recombining genetic material involves the exchange of material between homologous chromosomes. This process is called crossing over and occurs during meiosis. Crossing over involves pairing between homologous chromosomes, breakage and reunion of non-sister chromatids. This is illustrated in Figure 7.2.

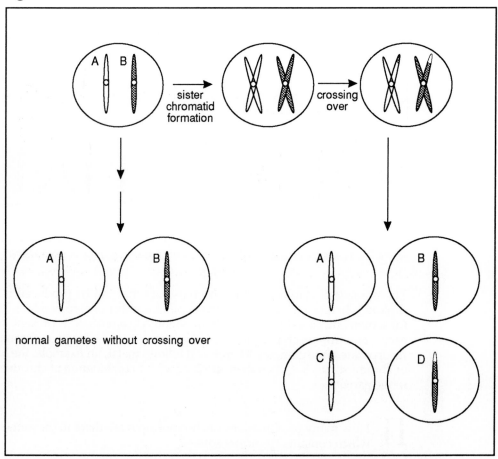

Figure 7.2 Crossing over (a single homologous pair of chromosomes are illustrated for simplicity).

Think about a sequence (a-e) of genes along a chromosome.

Normally we would expect a-e to be linked together since they are carried on the same chromosome. Because of crossing over they could become separated, eg:

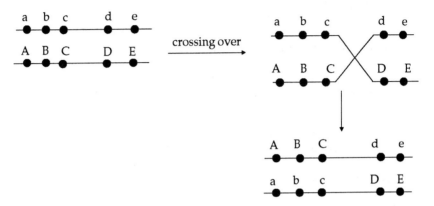

In the case illustrated, genes abc have been separated from d and e. Note that for simplicity each chromosome is depicted as a single thread (ie the two sister chromatids are not shown).

Ⅱ   Choose from the following the most likely result of crossing over in the case of the chromosome illustrated above:

• a separated from bcde;

• ab separated from cde;

• abc separated from de;

• abcd separated from e.

Your answer should be abc from de. The closer genes are together on the chromosome the less likely it is that a cross-over will occur between them (ie the more closely they are linked). Consequently, a will be most often separated from e, since these are furthest apart.

Ⅱ   If we began with a diploid cell containing the homologous chromosomes:

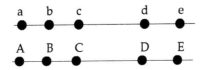

Which of the following combination of genes would we expect to find in greatest number amongst the progeny: Abcde; ABcde; ABCde; ABCDe?

the more closely genes are linked, the less frequently crossing over occurs between them

Most of the progeny would contain either ABCDE or abcde (ie they would contain chromosomes that have not crossed over), amongst those that did contain chromosomes that had crossed over we would expect the ABCde and abcDE combinations to be most common because the distance between genes c-d is greatest. In other words, we can use the recombination frequency to determine how closely genes are linked. Examine the illustrated chromosomes again. Note that all single cross-overs of these chromosomes will separate a and e (ie they are far apart) whereas very few will separate a and b (ie they are close together).

In mapping genes we have so far learnt how in principle we can assign genes to particular chromosomes and how we can determine the relationship (linkage) of these genes to others on the same chromosomes. Let us now put these principles into practice.

7.2.1 Two point crosses

In order to recognise the phenotypic expression of a gene one needs to have strains that carry different alleles of the gene in question. Usually, a gene can be identified when a mutation causes the occurrence of a recessive allele (absence of the trait). We will now analyse the expected segregation patterns for linked and unlinked genes. It is conventional to represent dominant alleles with capital letters, and recessive alleles in lower case.

segregation points

Suppose we have a pea variety with coloured flowers (AA) and wrinkled seeds (bb) and a variety with white flowers (aa) and round seeds (BB). The F_1 cross between the two varieties will give rise to a progeny, heterozygous for the two genes and showing the phenotype of the dominant allele: genotype AaBb, phenotype coloured flowers and round seeds. In this case the two genes are in the repulsion phase in both varieties (dominant alleles of one gene, combined with recessive alleles of the other).

repulsion phase

coupling phase

In the other situation, called the coupling phase, one strain is homozygous dominant for both genes, whereas the other is homozygous recessive for both. The consequences of these two possible situations for linkage analysis are illustrated below. Follow the sequence through.

Linkage analysis for two genes in F_2 progenies

	repulsion		coupling	
P (Parents)	coloured, wrinkled x white, round		coloured, round x white, wrinkled	
	↓		↓	
	AAbb x aaBB		AABB x x aabb	
gametes	Ab aB		AB ab	

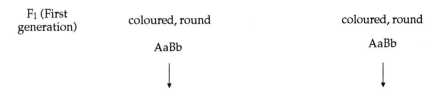

F₁ (First generation) coloured, round   coloured, round

AaBb   AaBb

If we assume no linkage between A and B (or a and B), then the gametes that can be produced from these are AB, aB, Ab and ab. They can combine in the following combinations:

♀ ♂	AB	Ab	aB	ab
AB	AABB	AABb	AaBB	AaBb
Ab	AABb	AAbb	AaBb	Aabb
aB	AaBB	AaBb	aaBB	aaBb
ab	AaBb	Aabb	aaBb	aabb

Those which carry an A allele will be coloured and those which carry a B allele will be round. Those which carry the two recessive alleles aa will be white, and those which carry bb will be wrinkled.

∏ We remind you that the F₂ generation is produced by selfing the F₁ generation. What proportions of the F₂ progeny will have the following phenotypes? (Use the table of gene combinations given above).

- coloured and round;

- coloured and wrinkled;

- white and round;

- white and wrinkled.

Your answer should be 9:3:3:1. Thus, of the combinations given: 9 carried A and B alleles and would produce coloured and round phenotype; 3 carried at least one A allele and bb (coloured and wrinkled); 3 carried aa and at least one B allele (white and round) and 1 carried aa and bb (white and wrinkled).

In conclusion, therefore we expect two unlinked genes to segregate in a 9:3:3:1 ratio in the F₂ progeny.

Let us examine the same genotype, but in this case let us use the example in which the genes A and B are tightly linked.

We begin with the same parents:

	repulsion	coupling
P	AAbb x aaBB	AABB x aabb
gametes	Ab aB	AB ab
F₁	AaBb	AaBb
	coloured round	coloured round

In this case however the gametes would be either Ab and aB (repulsion) or AB and ab (coupling) because the genes are linked. Therefore when we cross this F₁ generation the possible combination of genes would be.

gametes: Ab ♀ aB   gametes: AB ♀ ab

♂	Ab	AAbb	AaBb
	aB	AaBb	aaBB

♂	AB	AABB	AaBb
	ab	AaBb	aabb

∏ What proportions of the offspring would have the phenotypes described below in repulsion phase and coupling phase crosses?

	repulsion	coupling
coloured and wrinkled (c + w)		
coloured and round (c + r)		
white and round (w + r)		
white and wrinkled (w + w)		

You should have obtained a ratio of 1:2:1 c+w;c+r;w+r) for the repulsion phase and 3:1 (c+r; w+w) for the coupling phase situation. Thus for repulsion phase crosses:

genotype	number	phenotype
AAbb	1	coloured, wrinkled
AaBb	2	coloured, round
aaBB	1	white, round

and for coupling phase crosses:

A-B-	3	coloured, round
aabb	1	white, wrinkled

In summary, two unlinked genes are expected to segregate in a 9:3:3:1 ratio in the F₂ progeny; two completely linked genes will segregate in a 1:2:1 ratio in a repulsion phase cross and in a 3:1 ratio in a coupling phase cross.

We will now analyse the situation in which two genes are not completely linked but show a certain degree of crossing over, resulting in recombination between them. This is the situation when two genes are on the same chromosomes but are some distance apart. The distance between the two genes influences the frequency with which crossing over takes place. There are many factors, besides distance, that can increase or decrease the crossing over frequency. For these reasons the genetic distance between genes does not always reflect the physical distance between genes. If there is a high frequency of recombination between two genes, the physical distance can be relatively small. Crossing over frequency can be very low on the other hand, while the genes analysed can be located at quite some distance from each other on the different chromosome arms. Thus linkage analysis only tells us something about the order in which the genes are localised.

genetic distance and physical distance

If two gene pairs are not completely linked the process of crossing over during gametogenesis will lead to the formation of recombinant gametes. Suppose that genes A and B show a crossing over frequency of 0.2 (crossing over occurs in 20% of the cases).

We begin with the parents and the formation of gametes. Remember that A (or a) is on the same chromosome as B (or b). Thus:

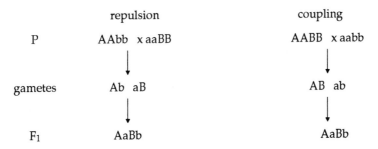

	repulsion	coupling
P	AAbb x aaBB	AABB x aabb
gametes	Ab aB	AB ab
F_1	AaBb	AaBb

The gametes that are formed from the F_1 generation will be predominantly Ab and aB for repulsion phase and AB and ab for coupling phase.

However, because of crossing over, a portion of new chromosomes are produced carrying the genotypes AB or ab and Ab or aB. If the frequency of crossing over is 20%, one fifth of the gametes will be represented by these genotypes.

Thus we can represent the gametes and the combination of genes in the F_2 generation as:

	Repulsion		new classes			Coupling		new classes	
gametes: ♀	4Ab	4aB	1AB	1ab	gametes ♀	4AB	4ab	1Ab	1aB
♂					♂				
4Ab	16AAbb	16AaBb	4AABb	4Aabb	4AB	16AABB	16AaBb	4AABb	4AaBB
4aB	16AaBb	16aaBB	4AaBB	4aaBb	4ab	16AaBb	16aabb	4Aabb	4aaBb
1AB	4AABb	4AaBB	1AABB	1AaBb	1Ab	4AABb	4Aabb	1AAbb	1AaBb
1ab	4Aabb	4aaBb	1AaBb	1aabb	1aB	4AaBb	4aaBb	1AaBb	1aaBB

We can recognise the following numbers in each phenotype class:

	repulsion	coupling
• coloured and round	(A-B-) 51	(A-B-) 66
• coloured and wrinkled	(A-bb) 24	(A-bb) 9
• white and round	(aaB-) 24	(aaB-) 9
• white and wrinkled	(aabb) 1	(aabb) 16

∏ Which of the F$_2$ progeny can be recognised phenotypically as containing recombinant chromosomes? To help you to do this look at the table of genotype given for the F$_2$ progeny. Examine those which are products of the new classes (crossed over) chromosomes. Which of these can be phenotypically distinguished from the progeny of gametes with uncrossed chromosomes?

In the case of repulsion only one progeny plant (aabb) out of a hundred can be recognised phenotypically as containing recombinant chromosomes. In the case of coupling, 18 progeny plants (A-bb and aaB-) out of a hundred can be recognised as being recombinants. Since segregation ratios in progenies follow statistical rules, one can imagine that trying to obtain a good estimate of genetic distance would require the analysis of large to very large F$_2$ repulsion progenies as compared to F$_2$ coupling progenies. In fact F$_2$ progenies should be used only to determine whether or not two genes are linked.

SAQ 7.1

Assume that flower colour and seed form are both single gene characters. Each exists in two allelic forms (red and white flowers; round and wrinkled seeds). Red is dominant over white and round is dominant over wrinkled. We will represent flower colour genes by F (or f) and seed shape genes by S (or s).

1) What will be the genetic composition (genotype) of a white flowered, wrinkled seed plant?

2) What will be the genetic combination(s) which will specify red flowers and round seed?

3) A homozygous red flowered and round seed plant is crossed with a white flowered, wrinkled seed plant. The progeny (F$_1$ generation) is selfed to produce an F$_2$ generation. This generation contains plants with the following phenotypes, white and round and red and wrinkled in a ratio of 3:1.

   Are the genes linked? (Draw out a flow diagram like that shown in section 7.2.1).

4) In the cross described in 3, careful examination of the large number of F$_2$ progeny showed that they were not all white and round or red and wrinkled. three in every thousand were white and wrinkled and 3 were red and round:

   a) What can you conclude about the flower colour and seed shape genes?

   b) What would be your conclusion if about 20% of the F$_2$ progeny were white and wrinkled or red and round?

7.2.2 Two point test crosses

In order to get a more precise measure of linkage one can use a double recessive progeny plant (aabb) in a two point test cross.

In a test cross an F_1 between two strains is crossed with a strain, which is homozygous recessive for the two genes to be analysed for linkage.

In the example below, we have again used a 20% cross-over frequency.

Thus if we are trying to determine the genetic distance between a set of genes (ie A or a and B or b) we begin with the following matings.

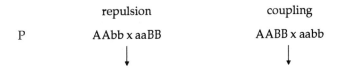

	repulsion	coupling
P	AAbb x aaBB	AABB x aabb

These will produce the F_1 progenies AaBb; AaBb.

These are then mated with a double recessive strain (aabb). Thus:

AaBb x aabb   AaBb x aabb

The gametes produced by the F_1 progeny will be:

4Ab,   4aB,   1AB**,   1ab**   and   4AB,   4ab,   1Ab**,   1aB**

**products of crossing over.

The gametes from the double recessive will be ab. Therefore the products of the crosses AaBb x aabb will be:

	4Ab	4aB	1AB	1ab
ab	4Aabb	4aaBb	1AaBb**	1aabb**

	4AB	4ab	1AB	1ab
ab	4AaBb	4aabb	1Aabb**	1aaBb**

Π Which of these products will have the phenotypes of the original parents?

The answer is: 4Aabb and 4aaBb   4AaBb and 4aabb

Π Which will have different combination of traits to the parents.

They are: 1AaBb** 1aabb**   1Aabb** 1aaBb**

These novel combination of phenotype characters are all products of recombinant chromosomes.

In other words, any combination of recombinant chromosomes can be identified from their phenotypic features.

There is no difference in calculating the crossing over frequency in the case of repulsion or coupling. In both cases all recombinants can be recognised. The crossing over frequency is simply calculated by dividing the total number of recombinants by the total number of progeny plants. The larger the test cross population will be, the better the estimate of the genetic distance.

With the analysis of two point crosses, one can get an idea whether two genes are linked or not. To determine the linear order of a number of genes, one can either carry out all possible two point test crosses, or three (multiple) point crosses.

| **SAQ 7.2** | We will again use our white and red flowered and round and wrinkled seed as an example. Remember that red is dominant (F) over white (f) and round (S) is dominant over wrinkled (s). |

We crossed a homozygous red flowered, wrinkled seed plant with a homozygous white flowered, round seed plant to produce a red flowered, round seed F1 progeny. This progeny was crossed with a white flowered, wrinkled seed plant.

The result was that the progeny of this cross contained:

340 red flowered wrinkled seed plants;

336 white flowered round seed plants;

24 white flowered, wrinkled seed plants;

22 red flowered round seed plants.

From this data calculate the crossing over frequency. (We suggest you draw out a sequence showing the genomes of parents and gametes and progeny to help you answer this question).

We have so far demonstrated how to determine the genetic linkage between two genes. But how do we put genes into a sequence along a chromosome? Suppose we have three genes A, B and C. Using the type of experiment described above we could determine the linkage between A and B and B and C. Suppose it was found that 20% crossing over occurred between genes A and B and 10% crossing over occurred between B and C. There are two possibilities for the crossing over (as %) between A and C:

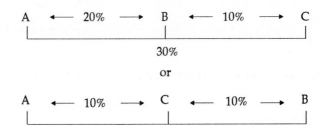

and you can choose between both possibilities by carrying out a linkage analysis for the two genes A and C. A more reliable and convenient test would be the analysis of a three point test cross.

7.2.3 Three point test crosses

In a three point test cross, two strains are used that have opposite genotypes for three genes.

For example AABBCC x aabbcc.

The F_1 progeny will therefore have the genotype AaBbCc.

This progeny is crossed with the triple recessive, thus: AaBbCc x aabbcc

If the genes are on separate chromosomes they will sort independently; thus the heterozygous F_1 plants will produce eight (ie 2^3) different kinds of gametes (eg ABC, ABc, Abc, abc, aBC, abC, aBc, AbC). These will occur with equal frequency. If, on the other hand, the genes occur on the same chromosomes the different gametes will occur in unequal frequencies.

Thus, if no crossing over occurs then gametes ABC and abc will be produced.

Π If the order of genes is ABC, what will the genotype of the gametes be if there was a cross-over between A and B? Use a drawing like this:

A	B	C		a	B	C
a	b	c	→	A	b	c

The gametes would thus be aBC and Abc. Use a similar drawing to work out the genotype of the gametes if a cross-over occurs between B and C. You should come to the conclusion that the gametes would be ABc and abC.

Π How could gametes with the genotypes AbC and aBc be produced? (Use the diagram above to help you)

The answer is the double cross-over thus:-

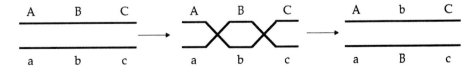

Now let us return to the cross of the F_1 progeny with the triple recessive plants ie AaBbCc x aabbcc.

Thus we can get the gametes:

F1 gametes		Products	
ABC abc		AaBbCc aabbCc	uncrossed chromosomes
ABc abC	mated with abc gametes	AaBbcc aabbCc	single cross-over between B and C
aBC Abc		aaBbCc Aabbcc	single cross-over between A and B
AbC aBc		AabbCc aaBbcc	double cross-over

Suppose the gene order in the chromosome is A-B-C and that the crossing over frequency between A and B is 10% and between B and C, 20%. You can imagine that the parental gametes will be present in the highest frequency. Ten percent of the test cross progeny will show recombination between A and B, whereas 20% will show recombination between B and C. Crucial to the determination of the linear order of the genes is the occurrence of the double cross-overs: 20% of the A-B recombinants will also show recombination for the B-C interval. Thus, we can expect that in this case 2% of the total progeny will be the result of a double recombination event. The two reciprocal double cross-over types (AbC and aBc) will occur in the lowest frequency. The order of the genes A, B and C can thus be deduced. We can draw this as:

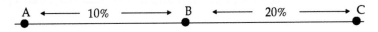

Let us examine an example. In a three point test cross the following data were obtained:

Genotype	Number
AaBbCc	345
aabbcc	363
aaBbCc	52
Aabbcc	43
AaBbcc	85
aabbCc	97
AabbCc	7
aaBbcc	8

The total progeny is 1000. Of these 52 + 43 + 7 + 8 = 110 show recombination between A and B. The cross-over frequency is therefore 110/1000 = 0.11 or 11%. Likewise, the cross-over frequency for the genes B and C is 85 + 97 + 7 + 8 = 197/1000 = 0.197. The expected double cross-over frequency is 11% of 19.7 or 0.11 x 0.197 = 0.018 or 1.8%. The observed frequency is 7 + 8/1000 = 0.015 or 1.5%.

We can draw the order thus:

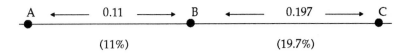

We draw to your attention that usually fewer double cross-overs are observed than expected. Apparently crossing over in one area influences the chance for crossing over in a nearby area of the chromosome. This process is called interference.

Now attempt SAQ 7.3.

SAQ 7.3

In a similar three point test cross to that described above, the following data was obtained.

Genotype of F_2 progeny	Number
AaBbCc	690
aabbcc	726
aaBbCc	170
Aabbcc	194
AaBbcc	14
aabbCc	16
AabbCc	86
aaBbcc	104

Use this data to draw a map of the genes A, B and C on the chromosome.

We should also draw your attention to trisomics and telotrisomics and their use in genetic mapping.

Trisomics

definition of a trisomic

Normal plants all have 2 copies of a specific chromosome, trisomics have one copy extra. Such a trisomic will produce disomic (AA) and monosomic (A) gametes (they will be monosomic for the other chromosomes). In principle, we can use this system to identify genes on the extra chromosome.

For *Petunia hybrida*, which has 2n = 14 chromosomes, a trisomic will have 15 chromosomes. Since there are 7 chromosome pairs, there will be 7 different trisomics, each exhibiting its own specific phenotype. In the case of *Petunia* you would have to check all seven possible trisomics in order to be able to unequivocally assign a gene to a specific chromosome.

Telotrisomics

definition of telotrisomics

Telotrisomics have only one arm of a chromosome in excess. Thus where use of a trisomic will provide information on the chromosome that contains a gene, a telotrisomic can add the information about which arm of the chromosome holds the gene.

7.2.4 Summary of cross-over approaches to gene mapping

We have learnt that by analysing the progeny of crosses between F_1 progeny and homozygous recessive strains or by crossing F_1 with F_1 progeny, we can determine whether or not genes are linked. We can use the same kind of analysis to determine how close genes are to each other and we can arrange them in a linear order.

Using this kind of analysis, genetic maps have been constructed during the past 60-70 years for a number of species like *Zea mays*, tomato, *Arabiodopsis thaliana* and *Petunia hybrida*. This is, however, a rather slow process.

During the past ten years, a very powerful technique has been developed that has greatly speeded up gene mapping. This technique is called Restriction Fragment Length Polymorphism (RFLP) analysis. We will examine this technique in the next section.

7.3 Restriction Fragment Length Polymorphism (RFLP) analysis

Restriction enzymes have the ability to cut DNA at specific sites. As a result, DNA is segmented into discrete parts. Restriction sites consist of specific nucleotide sequences. If a mutation occurs in such a sequence, the restriction site will be changed and the restriction enzyme will no longer recognise this new site for cutting. On the other hand such a sequence might be induced by a mutation. For example, the enzyme *Eco*R1 cuts at the six nucleotide sequence GAATTC. A mutation leading to the replacement of the second nucleotide by a T will destroy this particular site and thus *Eco*R1 will no longer be able to cut this sequence. Likewise a change from GATTTC to GAATTC will induce an *Eco*R1 site. The place of a given restriction site, and thus the length of a given restriction fragment, can be changed through deletion, insertion or inversion. We have illustrated these changes in Figure 7.3.

∏ Examine Figure 7.3 carefully and list two ways in which the products of DNA hydrolysis by restriction enzymes may differ when the bases in the DNA are changed.

Base changes can give rise to a different number of fragments or, more usually, fragments of different sizes.

∏ How can we separate the DNA molecular fragments?

electrophoresis The most useful way is to make use of the charge on their phosphate groups and the differences in the size of the fragments, ie we can use electrophoresis. In electrophoresis, the mixture of DNA fragments are placed, in a buffer, at one end of a gel, usually made of polyacrylamide and an electric field applied across the gel. The negatively charged DNA fragments migrate twoards the anode. The rate at which they migrate depends upon their size.

∏ Look at Figure 7.3. If we carried out gel electrophoresis with each of the restriction enzyme fragments shown, which fragment would migrate the quickest in an electric field and which the slowest?

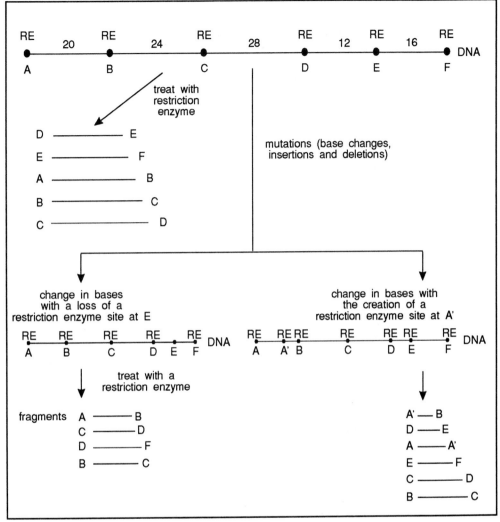

Figure 7.3 Restriction enzyme fragmentation of DNA and the consequences of mutation (RE = restriction enzyme site, numbers represent relative length of DNA sequences between restriction enzyme sites).

separation on the basis of size

You should have chosen the smallest fragment as the fastest migrator and vice versa. Consider the mixture of nucleotide chains (numbers represent the relative length of the nucleotide chain):

$$A \overset{20}{\rule{2cm}{0.4pt}} B \qquad B \overset{24}{\rule{3cm}{0.4pt}} C \qquad F \overset{16}{\rule{2cm}{0.4pt}} G$$

$$C \overset{28}{\rule{3cm}{0.4pt}} D \qquad E \overset{12}{\rule{1cm}{0.4pt}} F$$

If we put these at the top of a gel and apply an electric field, then they would separate out thus:

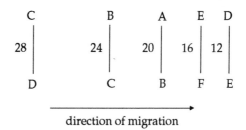

direction of migration

To detect these we can use a stain which reacts with the DNA. More usually, however, we would like to know which fragments contain a particular DNA sequence or gene. To detect these, we use a DNA probe. The probe contains a nucleotide sequence which is complementary to the gene sequence. Probes are usually produced using radioactively labelled nucleotides.

hybridisation

Southern blotting

Thus the procedure is to take the gel and blot this onto a filter. DNA from each band is adsorbed onto the filter. The DNA on the filter is then melted (separated into single strands) and incubated with the radioactive DNA probe. The probe DNA will stick (hybridise) to those fragments which contain complementary sequences. This technique is known as Southern blotting. We will discuss these probes in a later section.

If you are unfamiliar with the technique of Southern blotting, we would recommend the BIOTOL text "Techniques for Engineering Genes" which provides many of the technical details of this procedure.

Now let us apply this procedure to gene mapping.

Below is an autoradiogram developed from a Southern blot. For this, DNA was extracted from two homozygous parent strains (A and B) from the F_1 progeny of a cross between these parents and from back-crosses between these F_1 progeny and the parent strains. These DNA samples were treated with a restriction enzyme and the fragments separated by electrophoresis. Then using a DNA probe and the Southern blotting procedure, the DNA carrying nucleotide sequences complementary to the probe were detected.

A	B	F_1	A x F_1 or	B x F_1 or	Markers (mol wt kbp)
▬		▬	▬ ▬	▬	▬ 2
					▬ 4
	▬	▬	▬	▬ ▬	▬ 6

probes detect amount of gene present

The two parent lines A and B appear to each contain a single restriction enzyme fragment which carries the gene from which a probe was prepared. A cross between these two lines will show a hybridisation pattern harbouring both bands of the parental lines (see F_1 above), theoretically at half the intensity of the parental lines (both parents are homozygous, whereas the F_1 is heterozygous). In a test cross progeny of the F_1 with line A two types of banding pattern will appear. Some of the progeny will be homozygous for the A fragment, the others will be heterozygous. Likewise, in the test cross progeny of the F_1 with line B, both the homozygous type for the B fragment and the heterozygous type will occur amongst the progeny. In this way one can analyse for example 50 progeny plants for each test cross. Note that we usually describe the size of restriction enzyme fragments in terms of kilobasepairs (kbp).

After this analysis one can strip the filter (remove the probe) and re-hybridise the same filter to another probe.

∏ Consider the following example. What can you conclude about the linkage of gene A and gene B?

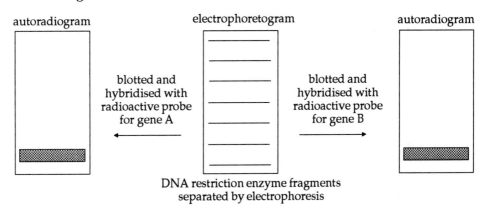

DNA restriction enzyme fragments
separated by electrophoresis

Genes A and B are presumably very close together because there is not a restriction enzyme site between them. If there had been, the probes would have hybridised to different fragments. (This is not however entirely waterproof: the two genes might be on equally sized fragments. These will coincide even without linkage. Thus we cannot be certain about linkage using DNA from a single plant).

∏ Examine a similar set from another experiment. What can you conclude about the linkage of gene A and gene B?

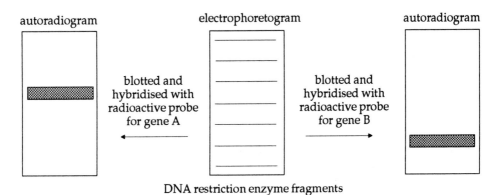

DNA restriction enzyme fragments
separated by electrophoresis

The only firm conclusion you can come to is that there is at least one restriction enzyme site between the two genes. There could be many more than one such site.

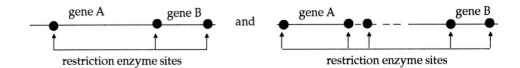

After hydrolysis, the fragments which carry genes A and B will each migrate separately during electrophoresis. If we use a different restriction enzyme we will get different fragments produced. Consider the sequence:

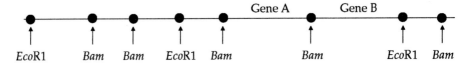

where *EcoR*1 and *Bam* are sites of action for two different restriction enzymes. If the DNA was treated with *EcoR*1, genes A and B would appear on the same fragment. If the DNA was treated with *Bam*, the genes would appear on different fragments.

This type of analysis, the so called Restriction Fragment Length Polymorphism (RFLP) analysis enables us to detect genes and determine the relationship of closely linked genes. RFLP analysis has found application in helping to determine segregation and linkage patterns.

RFLP map   For some species like corn, tomato and *Arabidopsis*, an extensive RFLP map containing hundreds of markers has become available over the past few years. The biggest advantage of this technique is the fact that we are no longer dependent on the availability of alleles of a specific gene, that show differences in expression. The RFLP technique is based on the occurrence of restriction site variation and the restriction sites can be outside of the sequence of the gene in question. We can even use undefined random cDNA fragments as a probe. The use of this kind of probe will be discussed in the next part of this chapter.

We have described the analysis of specific genes (and their mapping) through the analysis of the effect that different alleles have on the phenotype. Classical genetics started with the description of alleles by observing and analysing the different phenotypes they give rise to. Subsequently biochemical analysis elucidated the gene product: proteins and enzymes. We will describe how genes can be isolated and studied at the DNA level.

SAQ 7.4

Southern blotting was carried out, using two probes A and B on the DNA from four F$_2$ progeny. In each case, the same restriction enzyme was used. What can be concluded from the autoradiogram?

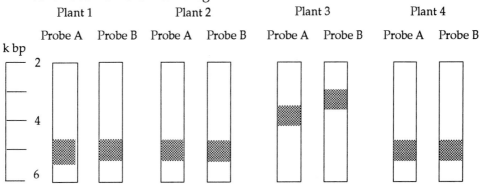

SAQ 7.5

After carrying out Southern blotting the following autoradiograms were obtained using probe A.

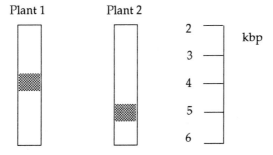

In a cross between plant 1 and plant 2, pollen from plant 2 was used to fertilise ovaries of plant 1. The F$_1$ progeny were grown up and analysed by Southern blotting. The following autoradiograms were produced from the F$_1$. Note all plants were diploids.

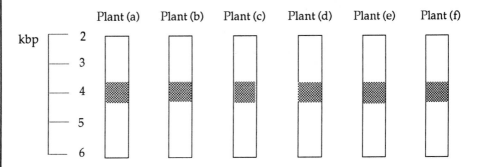

1) What can you conclude about the inheritance of gene A?

2) How would you test your conclusions?

7.4 Gene isolation

In the previous sections, we have considered the techniques that are available to us for mapping genes. In this section we deal with the techniques which have been developed for isolating genes. The technique which is used depends upon the gene of interest and how much is known about the gene and/or its product. Consider the following sequence.

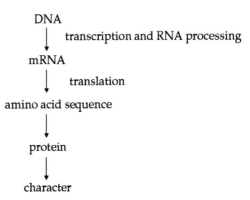

The genetic information in DNA, is first transcribed into an RNA sequence which is translated into an amino acid sequence giving rise to a protein which influences phenotypic character.

two basic strategies   Essentially what we are aiming to do here, is to attempt to isolate specific pieces of genetic information from DNA. We have two points of attack. We can either try to extract this information directly from the DNA, or indirectly using the information in messenger RNA (mRNA). It is usual to attempt to use the information in mRNA rather than that stored in DNA.

cDNA library   We will examine the reason for this later. We use the information in mRNA by producing a cDNA library. A cDNA library is a collection of copy DNA (cDNA) synthesised from a pool of mRNA. A cDNA library is produced in the manner described in the next section.

7.4.1 Construction of cDNA library

selective isolation of mRNA   In the first stage, a tissue is selected that expresses the gene of interest. The RNA is extracted from the tissue and the mRNA is purified from this total RNA. Use is made of the poly A (poly adenosine) 'tails' that are attached to mRNA. Such molecules hybridise with poly deoxyribose thymidine (TTTT) as shown in Figure 7.4.

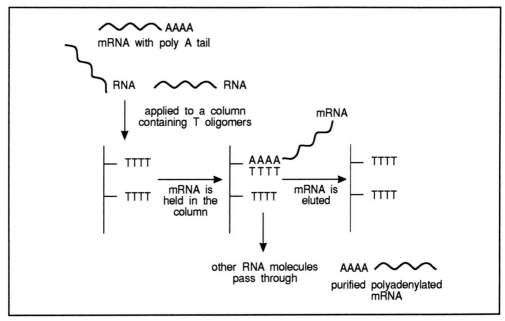

Figure 7.4 Purification of poly adenylated mRNA.

The isolated mRNA is then hybridised with a short, deoxyribose thymidine (dT) oligomer. Thus:

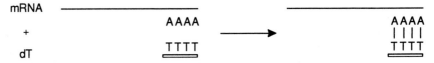

The oligomer of dT acts as a primer for the enzyme reverse transcriptase. This enzyme (from retrovirus) uses RNA as a template and synthesises DNA. Thus:

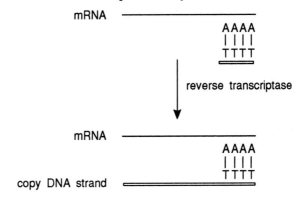

The RNA can be removed from these hybrid molecules by alkaline hydrolysis and then the second complementary strand of DNA is made using *E.coli* DNA polymerase I. Thus:

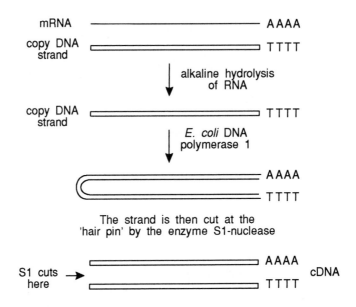

Using this technique, we have therefore converted the nucleotide sequence in the mRNA into a double stranded DNA sequence. We call such DNA, copy DNA or cDNA.

cloning of cDNA

The next stage in the process is to make detectable quantities of each of the cDNA species. This is done by a process called DNA-cloning using a plasmid or bacteriophage (phage) vector. You will recall that bacteriophages are viral-like particles that infect bacteria. We routinely use such bacterial hosts to make large quantities of the nucleotide sequences carried by bacteriophages. Here we will explain how this is achieved using a phage. Figure 7.5 summarises the steps involved. In outline the phage DNA is cut with a restriction enzyme, specific linker sequences are added to the ends of the cDNAs so that these ends fit exactly into the opened phage. Finally the cDNAs are covalently attached to the phage with the enzyme DNA ligase.

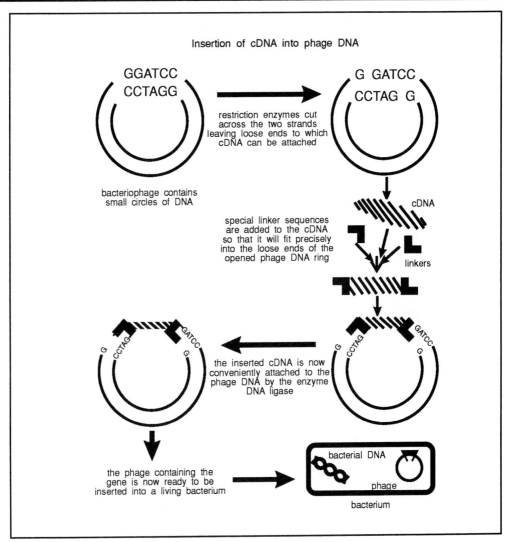

Figure 7.5 How cDNA is inserted into a bacteriophage.

Let us examine this a little more closely.

lambda vectors Often derivatives of the bacteriophage λ (lambda) are used. These derivatives have been specially produced so that they have special, single restriction enzyme sites. Some examples of these are the so called lambda vectors (eg λgt10, λgt11 etc).

When the lambda vector is cut with a restriction enzyme, a linear molecule with short single strands of DNA at each end is produced. In the example shown above these ends would have the structure:

GATCC ———————————————————— G
       G ———————————————————— CCTAG

To make these ends join up specifically with the cDNAs, we have to produce cDNA with the appropriate nucleotide sequences at its ends. We could for example add to each end of the cDNAs, the nucleotide sequence CTAG to produce sequences with the structure:

$$\text{CTAG} \overline{ \text{cDNA} } \text{CTAG}$$

If we now incubate the linear phage DNA with the cDNAs, then they will naturally align themselves as shown in Figure 7.6.

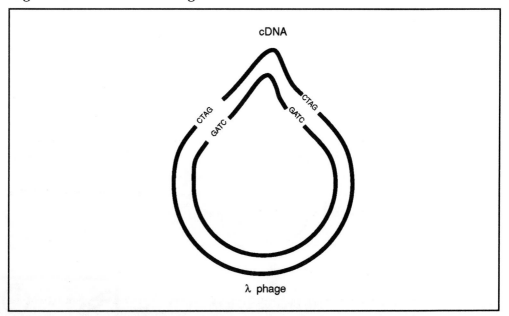

Figure 7.6 Insertion of cDNA into λ phage DNA.

A ligating enzyme will then join up the molecule. We can regard the GATC sequence added to the cDNAs as 'linking' sequences (they are usually referred to as linkers).

plaques   The phage DNA carrying the cDNA can now be packed into the λ phage heads and these introduced into an appropriate *E.coli* host strain which spread out as a lawn on solid media. The phage produce plaques (cleared zones in the lawn of the bacteria which grow up on the plates; Figure 7.7).

It is estimated that on average, if 20 000 plaques are produced using phage containing cDNA, then this will cover the complete set of genes being expressed in the tissue from which the mRNA was originally extracted. We call such a set a cDNA library.

selection of desired clones   The key question at this stage is to identify (select) which of the plaques were produced by phage containing the cDNA of interest. The strategy used depends on the nature of the vector used to clone the cDNA and on the nature of the gene product. Let us consider these each in turn.

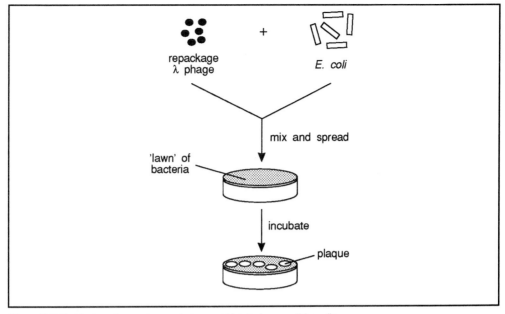

Figure 7.7 Producing phage plaques in lawns of bacteria on solid media.

7.4.2 Influence of vector on the screening of the cDNA library

Vectors fall into two main categories:

- expression vectors;

- non-expression vectors.

In expression vectors, the cDNA is inserted next to a promoter site. Thus, in such systems the cDNA is transcribed and translated to produce a protein (ie it is expressed). Thus:

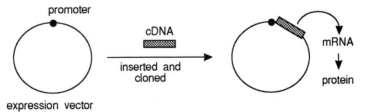

An example of an expression vector is λgt11.

In a non-expression vector, the cDNA is inserted at a site remote from a promoter. It is not transcribed. Thus:

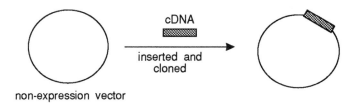

non-expression vector

An example of a non-expression vector is λgt10.

Thus in the case of an expression vector we have the option of either screening for cDNA or the protein produced from this cDNA. Clearly, in the case of a non-expression vector, we have to screen for the cDNA.

7.4.3 Screening the cDNA library

replica

Essentially we need to identify which of the plaques illustrated in Figure 7.7 represents which cDNA. We can press a filter into such a petri dish to make a replica of it.

plaque-lift

When a nitrocellulose or nylon filter is placed on the plate, part of each plaque will stick to the filter and can be lifted (plaque-lift). When these filters are treated so as to fix the DNA the result is a replica of the cDNA library on a filter that can be screened by DNA-DNA hybridisation. This general biochemical technique employs the capacity of separated complementary DNA strands to re-anneal (in the appropriate buffer) as a stable double stranded form. Double stranded DNA fragments can be denatured by alkaline treatment and fixed on a solid support like a nitrocellulose or nylon filter. These filter replicas can be obtained from plaques. With radioactively labelled DNA fragments or RNA molecules (probes), one can detect very small amounts of homologous DNA. We will discuss the production of probes in the next section.

This technique is very similar to the Southern blotting technique described earlier to demonstrate DNA sequences in electrophoretically separated fragments (see Figure 7.3).

The sequence for identifying particular cDNA clones in a series of plaques is illustrated in Figure 7.8.

The specificity of the hybridisation of the probe to the target DNA is influenced by the use of high salt buffers and the incubation temperature. Under certain conditions one can even have hybrids between DNA strands that are not completely homologous (low stringency hybridisation). This is essential when the homology between the probe and the target gene has not been established, but undesirable when one is looking for one specific DNA or RNA sequence (high stringency hybridisation).

Note that the techniques of DNA/DNA hybridisation on 'blotted' DNA was named Southern blotting after its inventor Ed Southern. We can also determine RNA on filters or in gels using an analogous method. This has been (facetiously) named Northern blotting. There is also a technique called Western blotting for detecting specific proteins. This is relevant when we produce a cDNA library in an expression vector.

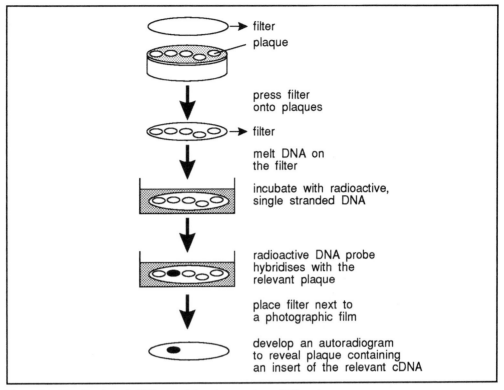

Figure 7.8 Identification of cDNA containing clones using a DNA probe.

We have illustrated what is occurring during the hybridisation with the probe DNA in Figure 7.9.

7.4.4 Constructing oligonucleotide probes

With the development of protein sequencing techniques and DNA synthesisers the following approach has become available. The microsequencer can determine the amino acid sequence of a peptide from only microgram quantities of purified protein. This amino acid sequence can then be used to synthesise the putative DNA sequence coding for these amino acids. Because the genetic code allows several bases at the third position of a triplet (codon degeneration), mixed oligonucleotides will have to be synthesised to assure that the correct DNA sequence is present.

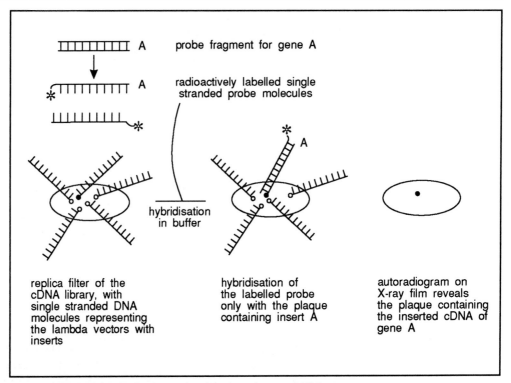

Figure 7.9 DNA/DNA hybridisation to reveal the homologous cDNA insert.

Π   Let us assume that we know that the following amino acid sequence is present in a gene of interest.

met val thr val glu glu tyr trp lys

We know that the codons for these amino acids are:

- met - ATG;

- val - GTG or GTC or GTA or GTT;

- thr - ACA or ACC or ACT;

- glu - GAG or GAA;

- tyr - TAT or TAC;

- trp - TGG;

- lys - AAG or AAA.

How many different nucleotide sequences could code for the amino acid sequence described above? (Try and work this out before reading on).

Your answer should be 768! It is helpful to draw out the amino acid sequence and to put in the possible nucleotide sequences as illustrated.

met	val	thr	val	glu	glu	tyr	trp	lys	...	...	amino acid
ATG	GTG	ACA	GTG	GAG	GAG	TAT	TGG	AAG			oligonucleotide
	C	C	C	A	A	C		A			
	A		A								
	T	T	T								
1 x	4 x	3 x	4 x	2 x	2 x	2 x	1 x	2			$n = 768$ possibilities

For some amino acids, two, three or even four triplets are used, while for others there is only one (in this case met and trp). Every option for A, T, C or G increases the number of different oligonucleotides that need to be synthesised. Thus if two different triplets are available (eg as with tyr) this gives 2 options, if three different triplets are used (eg thr) then this gives 3 options. Now by examining the bottom line of our diagram you will see that, even with a short amino acid sequence, many different oligonucleotides may code for this sequence. To make certain we can identify the sequence in a cDNA library, we need to synthesise a mixture of these oligonucleotides. This is accomplished by an automated DNA synthesiser.

This synthetic oligonucleotide mixture can be radioactively labelled and used to screen replicas of a cDNA library by DNA-DNA hybridisation. The DNA of the recombinant phage with the insert representing the gene, hybridises with the correct oligonucleotide probe of the mixture while other recombinants do not hybridise. Autoradiography of the hybridised replica filters will reveal the plaques that bind radioactive probe (Figure 7.9). The hybridising plaque is then localised on the original plate, re-plated on a new plate and screened again. In this way the lambda (λ) recombinant containing the cDNA of interest can be purified.

7.4.5 Antibody screening of an expression cDNA library

antibody
screening

When no DNA or protein sequence data are available one can use an immunological approach to identify the cDNA of interest providing this is produced in an expression vector. The protein that is thought to be the product of the gene of interest is used to raise antibodies in a rabbit (or other animal). The specificity of these antibodies is determined by their reaction to a dilution series of the protein. These antibodies can be used for the 'immunodetection' of very small quantities of protein in a complex mixture. Therefore they can be applied to detect proteins produced by recombinant phages of a cDNA library (Figure 7.10). The protein encoded by the cDNA insert will only be expressed efficiently if the cDNA library is constructed in the appropriate vector (ie λgt11).

selectable
markers

With expression vectors like λgt11 we adopt a rather specialised procedure. The vector carries a selectable marker (in the case of λgt11 it is an ampicillin resistant gene). The vector is also lysogenic (ie it does not cause lysis of the bacterial host). Thus, by using an ampicillin sensitive host, those cells infected by the vector acquire ampicillin resistance and can be grown on ampicillin containing media. The next key stage is to identify the colonies of bacteria which contain the cDNA of interest. This is achieved in the following way.

The ampicillin resistant bacteria are incubated at an elevated temperature (for λgt11 this is 42°C). This induces production of the protein encoded by the cDNA. Replicas of the ampicillin colonies are made onto a filter. The filters are incubated with antibodies raised against the protein of interest. These will react with the appropriate protein(s) attached to the filter and will enable identification of the colonies which are producing the appropriate protein (Figure 7.10). Often the antibodies that are used are either radioactive or have fluorescent dyes attached to them to facilitate their detection on the filter. The colonies harbouring the vector containing the cDNA of interest can thus be identified.

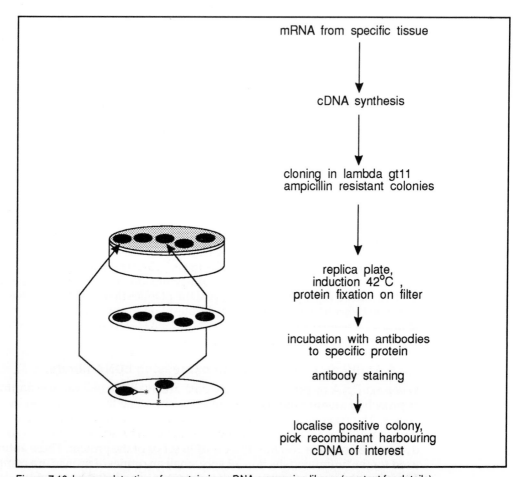

mRNA from specific tissue

cDNA synthesis

cloning in lambda gt11
ampicillin resistant colonies

replica plate,
induction 42°C ,
protein fixation on filter

incubation with antibodies
to specific protein

antibody staining

localise positive colony,
pick recombinant harbouring
cDNA of interest

Figure 7.10 Immunodetection of a protein in a cDNA expression library (see text for details).

7.5 Gene cloning by the polymerase chain reaction (PCR)

PCR

A technique developed in the last few years has changed the strategies for gene cloning drastically: the polymerase chain reaction (PCR). This method can be used to amplify specific DNA sequences from genomic DNA or cDNA mixtures to microgram quantities. No DNA library of a specific kind will have to be constructed and tedious screening procedures can be abolished.

In this section we will give a brief outline of the processes involved in the polymerase chain reaction. Greater technical detail can be found in the BIOTOL text "Techniques for Engineering Genes".

The method depends on the knowledge of the putative DNA sequence of the gene of interest. In the case of a sequenced protein, mixed oligonucleotide primers can be synthesised representing the amino terminal and the carboxy terminal amino acid sequence. We have labelled these A and B in Figure 7.11. Use this Figure to follow the description below.

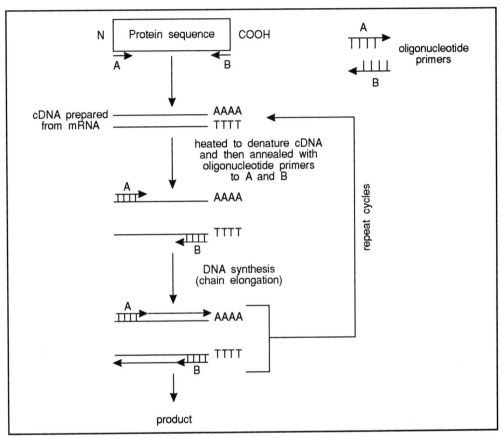

Figure 7.11 Outline of PCR amplification of a cDNA (see text for details).

We take the DNA sample (in the figure we have illustrated this as cDNA prepared from mRNA, but we could also have used genomic DNA which had been treated with a restriction enzyme) and it is melted (denatured) to produce single strands. The single stranded DNA is then incubated with the oligonucleotide primers (A and B). These will contain complimentary sequences to those at either end of the cDNA and will form a double stranded molecule as a result of base pairing. The process by which this is carried out is often called annealing or hybridisation. Thus we now have pieces of DNA which are mainly single stranded but which have short double stranded portions. We can now use *E. coli* DNA polymerase to complete the synthesis of the complementary strand (ie chain elongation). Thus we now have two copies of the DNA we began with.

After this step the strands can be separated thermally and the primers are allowed to hybridise again. Addition of DNA polymerase will initiate a new cycle of DNA synthesis, only now there are four copies of the target DNA strand. On each subsequent cycle this DNA fragment will be amplified.

automation
using thermally
stable DNA
polymerase

This 'polymerase chain reaction' should yield 2^n copies (n = number of cycles) of the DNA fragment, but after 25 cycles primers and nucleotides become limiting and the increase is no longer exponential. Still the PCR procedure can yield a microgram of product from a very complex DNA mixture with only femtograms (10^{-15}g) of target (gene of interest) DNA. With the isolation of the thermostable *Taq*-DNA polymerase from *Thermus aquaticus* automation of the PCR procedure became possible. This is because no enzyme has to be added after each thermal denaturation of the DNA strands, therefore the procedure can be executed by a computerised thermo-block. Amplified fragments can be purified and cloned into simple plasmid vectors or the DNA sequence can even be determined directly. Both methods described above rely on knowledge of the structure of the protein in order to isolate the gene. However, when a gene is isolated from one organism it can be used to identify homologous genes in other organisms, even when there is hardly any information on the enzyme. We will examine this in the next section.

SAQ 7.6

A protein is known to have the following amino acid sequence:

NH₂ met thr glu glu met trp .... met trp thr val trp trp COOH

The codons for these amino acids are:

met - ATG;

thr - ACA or ACT or ACC;

glu - GAG or GAA;

trp - TGG;

val - GTC or GTG or GTA or GTT.

Design oligonucleotide probes which could be used to act as primers for the production of genes coding for the protein of interest using the polymerase chain reaction.

7.6 Gene isolation with a heterologous probe

use of
heterologous
probes

The use of a heterologous DNA probe can be considered for related plant species, but even *E.coli* or yeast genes can be used for this purpose. The heterologous DNA fragment can be used as a radioactive probe to screen a cDNA - or genomic DNA library of the (plant) species of interest. This heterologous DNA fragment will bind only to the gene fragment that encodes a similar protein. Because there will be a difference in the DNA sequence between the genes of the two organisms the hybridisation conditions have to be considered carefully and should not be too stringent. When the DNA sequence of the heterologous gene fragment is known, primers can be designed for a PCR experiment.

7.7 cDNA or genomic DNA?

Most of the sections concerning the isolation of genes have focused on cDNA rather than genomic DNA.

∏ List the advantages and disadvantages of using cDNA? Use the following diagram of transcription and translation to help you.

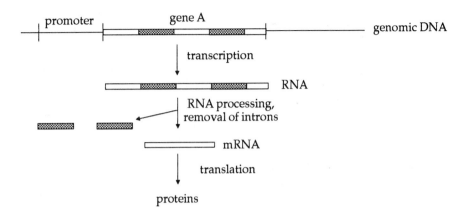

Copy DNA (cDNA) is simply a deoxyribose version of mRNA. In other words the nucleotides are in the correct order to be translated into a protein. Thus transfer of cDNA into a new host means that we transfer the information for a single gene/gene product. If we used genomic DNA this may contain introns (nucleotides sequences which are removed during RNA processing). There is no guarantee that the new host will have the mechanisms available for this processing.

Using mRNA as a starting point, means we are sure to produce the whole nucleotide sequence for a particular gene product. If we use genomic DNA cut into fragments, we have no guarantee that the enzyme will not cut part way along the gene. Thus:

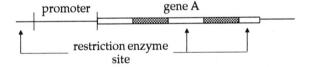

On the other hand the DNA fragment which is produced may carry the regulatory (promoter) gene portion which controls the expression of the gene. This promoter could be important if it was intended that the gene was to be expressed in a particular way. If cDNA is used, no such promoter would be isolated directly.

So far gene isolation has been considered for traits that were characterised as far as the protein level. Most of the plant traits have not however been characterised that thoroughly. They are specified only by a certain phenotype and a location on the genetic map of the plant. The isolation of these genes is more laborious and relies on a combination of genetic information and molecular techniques. We will examine these in the final section of this chapter.

7.8 Chromosome walking

Chromosome walking is one of the methods that can be applied to isolate gene fragments. When the trait is mapped on one of the plant chromosomes, random plant DNA fragments can be tested for linkage to the phenotype by the RFLP technique. In other words, we can trace the phenotype to long sequences of nucleotides isolated by restriction enzyme digestion of DNA. This sequence can be used as a starting point for what has been described as chromosome 'walking'.

Consider the chromosome carrying the gene (X) of interest.

We first isolated an RFLP fragment which is closely related to X. Thus:

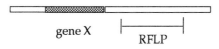

We now prepare more DNA fragments of the chromosome using other restriction enzymes. These will cut the DNA of the chromosome in a different way. These DNA fragments are cloned in a vector to produce a genomic DNA library. Using the Southern blotting technique and the RFLP fragment as a probe, we can identify fragments which carry common sequences to the RFLP. Thus:

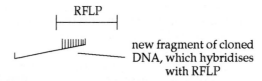

The clones that hybridise with the RFLP fragment can then be used as probes in Southern blots of genomic libraries prepared by other restriction enzymes to identify those carrying overlapping sequences.

contigs

By this technique we begin to construct a continuous sequence of nucleotides. These sequences are often referred to as contigs.

cosmid vectors

yeast artificial chromosmes

This process of identifying sequences of nucleotides is referred to as chromosome walking. We illustrate this in Figure 7.12. For this process a suitable genomic library should contain large fragments (10-30 kbp, cosmid vector) or very large fragments (50-200 kbp, yeast artificial chromosome, Yac vectors) of plant DNA. If you are unfamiliar with the terms cosmid vectors and yeast artificial chromosmes, these are two types of genetic vectors. Cosmids are plasmids containing special sites (cos sites) into which DNA can be readily inserted. Their special feature is that they enable fairly large fragments (10-30 kbp) of DNA to be cloned. The yeast artificial chromosome also enables DNA fragments to be cloned, but in this case, very large fragments (up to 200 kbp) can be used. Further details of these cloning vectors can be obtained from the BIOTOL text "Techniques for Engineering Genes".

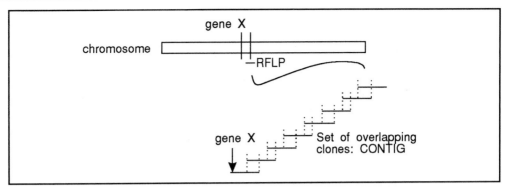

Figure 7.12 Chromosome walking.

From the position of a RFLP fragment linked to the gene X, overlapping genomic DNA fragments are selected by hybridisation. These fragments are linked together into a contig (set of linked fragments). These contigs can be linked up to form a physical map of the chromosome. One of the genomic DNA clones will contain gene X.

∏ Let us carry out this process as an example. An RFLP fragment prepared by the restriction enzyme *EcoR1* is found to be linked with the phenotypic character of interest. This fragment will hybridise with DNA from plaque number 84 prepared by treating the chromosomal DNA with the restriction enzyme *Bam* I. The genomic DNA from plaque number 84 will also hybridise with a cloned genomic DNA sample prepared by treating the chromosome with restriction enzyme *Hind* III. From this information draw a restriction enzyme map of the chromosome in the region linked with the phenotypic character of interest.

You should have argued as follows. The RFLP fragment was prepared by *EcoR1* treatment. From this we could draw:

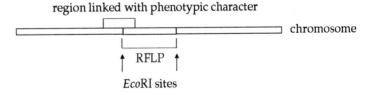

The *Bam*I sites overlap with part of this RFLP fragment. Therefore we could draw:

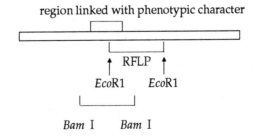

This latter fragment overlaps with a *Hind* III fragment. Thus:

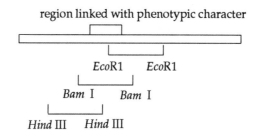

Thus we can draw the map of the region of interest as:

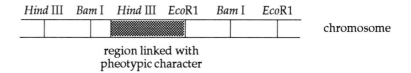

This example is rather simplified, but it will have enabled you to understand the process. Usually this physical map of the chromosome allows the selection of an area of approximately 500-1000 kbp that should contain the gene of interest. In the case above, partial digestion with *EcoR1* and *BamI* could produce fragments which contain the gene of interest.

The next step towards the isolation of the gene would be identification of transcribed regions of this physical map by probing the collection of genomic DNA fragments with labelled poly A-mRNA. When mRNA from plants mutated in the trait and from wild type plants is used, clones can be selected that hybridise only with wild type mRNA. These are likely candidates for the gene of interest. Definite proof that a DNA fragment is responsible for a certain trait needs more experiments and will be discussed at the end of this chapter.

7.9 Transposon tagging

transposon   While the former method relies on molecular technology to scan across the genome, transposon tagging utilises a piece of mobile DNA to randomly mutagenise the plant genome. Like most other organisms, plants contain mobile DNA fragments. These are called transposons. These transposable elements can excise from the DNA and integrate randomly into the chromosome. When the integration takes place in the structural or regulatory part of a gene this will result in a mutant allele of the gene. This is illustrated in Figure 7.13. In the progeny of a selfing of such a mutated plant, the transposon inactivated gene will be homozygous in one quarter of the progeny. Characteristic for these transposon induced mutations is the unstable phenotype that is caused by the frequent excision of the transposon from the target gene, thus restoring (partial or complete) gene function. The use of a transposon as a mutagenic agent is obvious, but more important is the molecular implication of the mutation. The mutated gene is 'tagged' by the transposon. If the DNA sequence of the transposon is known, it can be used as a probe. Using the transposon as a probe, one can clone the mutated gene from a genomic library, as the DNA fragment flanking the transposon.

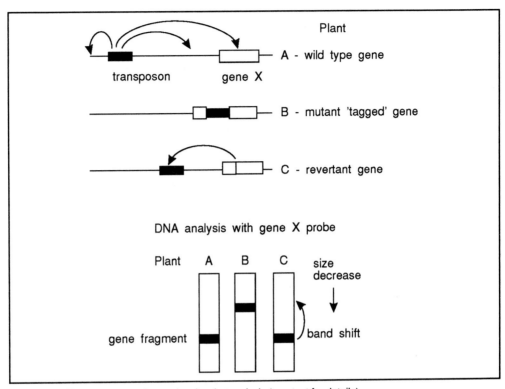

Figure 7.13 Transposon tagging and molecular analysis (see text for details).

Because multiple copies of the transposon can be present in the genome, the definite proof for the identity of the gene will have to come from further DNA analysis of wild type, mutant and revertant plants. The mutant's allele of the gene will differ in size from the wild type and the revertant allele because the transposon has been inserted. DNA analysis can reveal these differences (band shift) when we use the putative gene fragment as a probe.

band shift

7.10 Differential screening of cDNA libraries to obtain process specific genes

differential
screening

All the strategies described above aim at the isolation of a specific gene. However, a strategy has been developed to clone sets of genes involved in certain specific processes (eg pathogen attack, heatshock, a metabolic pathway expressed in a specific tissue, flowers, tubes, seeds, fruits, roots etc). This strategy is called differential screening. It relies on the specific expression of genes, induced by certain processes, or genes expressed in a tissue specific manner. The cDNA library made from mRNA extracted from induced tissue or the desired organ is probed with two sets of radiolabelled mRNA (or cDNA copies). For example, one from the non-induced tissue and one from the induced tissue, or it is probed with leaf mRNA and organ specific mRNA. This is illustrated in Figure 7.14.

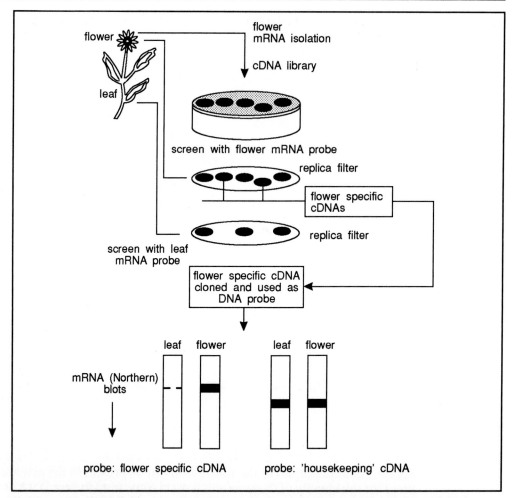

Figure 7.14 Simplified differential screening as a method to isolate flower specific cDNAs. A cDNA library is constructed from flower mRNA. Replica filters of this library are screened with radioactively labelled flower mRNA (or cDNA) and a duplicate is screened with labelled leaf mRNA. Flower specific cDNAs are detected as spots that only light up with the flower mRNA probe. Other spots that also hybridise to leaf mRNA represent 'housekeeping' genes. To check the specificity of the cDNA, individual clones are purified and used as a probe on mRNA blots from leaf and flower tissue. A specific flower cDNA will only hybridise to a mRNA band in the flower preparation.

housekeeping genes

differentially expressed genes

Different spots will light up on the replica of the cDNA libary but also identical spots. The identical spots represent the more general 'housekeeping' genes, expressed in both types of tissue, while the specific spots can be identified as differentially in expressed genes. These cDNAs are then purified and used as probes on RNA blots to confirm their specific expression. In this way tissue or process specific genes can be identified without knowing exactly what their function is. The selected cDNAs may be used to study the process in more detail and functions can be assigned to these cDNA fragments. One of the tools to characterise the gene function is gene transfer. Genes cloned by differential screening may not lead directly to a trait but can be a valuable pool of information for research on a specific plant process.

SAQ 7.7	The drawings below are related to identifying root specific genes. A number of plaques, produced by non-expression vectors carrying cDNA derived from root mRNA have been produced on a lawn of *E.coli*.

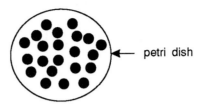

← petri dish

These have been replicated onto filters and incubated with radioactively labelled mRNA from leaves, roots and flowers. Autoradiograms developed from these filters are shown below. Use these to identify plaques which carry cDNA for root specific genes.

challenged with   challenged with   challenged with
leaf mRNA   root mRNA   flower mRNA

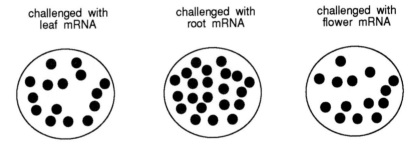

7.11 Linking the gene fragment to the trait: definite proof

To prove that we have definitely cloned the gene responsible for the trait we are examining several experimental tools are available.

7.11.1 DNA sequence determination

From the nucleotide sequence of the cloned DNA, the amino acid sequence and the molecular weight of the protein encoded by the cDNA can be deduced. When these data fit the available information of the biochemical characterisation, substantial evidence has been obtained to prove the identity of the cloned gene fragment. Sequence analysis can further be used to determine whether the genomic DNA of the gene contains introns. These can be localised by lining up the cDNA sequence and the genomic DNA sequence. Genomic DNA fragments, not present in the cDNA are introns that are spliced out during mRNA processing. The cDNA molecules from this mRNA will not contain this fragment.

DNA sequence determination of the genomic DNA fragment will also give information about the regulatory sequences in the promoter of the gene.

In some cases a single cDNA species will hybridise with more than one genomic DNA fragment. This indicates there are several genes containing similar nucleotide sequences. We could call these a gene family. Members of gene families can be distinguished by examining details of the promoter region and 3' tail of the genes or on the arrangement of introns (sequences of nucleotides not present in mRNA).

SAQ 7.8

We have used cDNA to explore gene structure in a particular plant. The diagrams below are stylised electron micrographs of hybrids of genomic DNA restriction enzyme fragments and a cDNA. Clear strands are genomic DNA, shaded strands are cDNA.

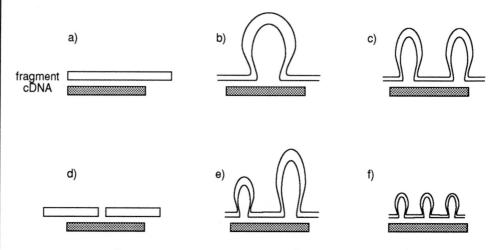

1) Which of the genomic DNA restriction enzyme fragments contain introns?

2) How many members are there in the family of genes related to that of the cDNA?

3) Which member of the family of genes has a restriction enzyme site in the middle of the gene?

7.11.2 *In vitro* translation

In vitro translation of mRNA molecules hybridising to the cDNA can be used to characterise the protein coded by these homologous mRNA molecules. This *in vitro* synthesised protein can be compared with the actual protein linked to the trait by gel electrophoresis and immunodetection experiments. The mRNA homologous to the gene fragment is 'fished' out of the total mRNA population by hybridisation to the single stranded cDNA molecule that has been immobilised on a filter. After this very specific purification, the homologous mRNA is eluted from the filter and translated by an *in vitro* system (eg by a rabbit reticulocyte or *Xenopus* oocyte extracts). If one of the amino acids present in the translation mixture is radioactively labelled with 35S (eg cysteine or methionine), the synthesised proteins will be radioactive as well. Radioactively labelled protein can be visualised by gel electrophoresis and autoradiography. In this way the product can be characterised and compared with the protein assigned to the trait. This characterisation can be completed by immunodetection with a specific antiserum.

7.11.3 Genetic analysis

The cloned cDNA fragment can be used as an RFLP probe to genetically map the gene on the chromosome. When this fragment is mapped on the same co-ordinate as the trait it is likely that we have cloned the gene of interest.

When the gene does not map on the same co-ordinate as the trait we have to consider that we are dealing with a gene family of which we have used a member other than the one involved in the selected trait. In that case the genomic DNA fragments encoding the different members of the gene family will have to be identified. The member of the gene family that is linked to the trait can then be assigned the gene of interest.

7.11.4 Gene transfer experiments

For many dicotyledonous plants the introduction of foreign genes has become possible. It is therefore possible to design a gene transfer experiment to assign a function to the cDNA or the genomic DNA fragment that has been isolated. There are two basic strategies to be distinguished:

- Complementation of mutant plants with the wild type gene isolated either as a cDNA or a genomic clone.

When the introduced gene restores the wild type phenotype partially or completely, the definite proof for the identification of the gene has been delivered. When a genomic DNA fragment is used it can be envisaged that the correct regulatory elements will be present on the fragment to be transferred. However when a cDNA fragment has to be tested in this way it will have to be equipped with the promoter and polyadenylation signal to ensure expression in tissue that exhibits the trait of interest. Although this can be a problem for very specific tissues, tissue specific promoters of various genes have been characterised and can be used for this purpose. These specific regulatory DNA fragments can be fused to the cDNA by DNA-recombinant technology and a reconstructed gene can be introduced into the mutant plant. Complementation of a mutant, resulting in the wild type phenotype gives a straightforward answer but when no complementation is visible in several independently transformed plants one cannot decide immediately that the gene fragment tested does not correlate with the trait. One has to confirm that the fragment is expressed in the transgenic plant, both at the mRNA level and the protein level.

- The second strategy is the use of an introduced gene in the antisense orientation in order to inhibit the wild type expression of the trait.

Let us explain this is a little more detail.

∏ Consider a DNA sequence and its transcription.

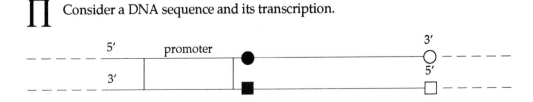

From your knowledge about gene expression, where will the transcription of gene A begin and which strand will be transcribed?

You should know that RNA synthesis will begin at the promoter that it is synthesised against the 3' - 5' strand of DNA. Thus:

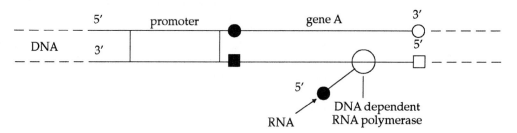

What would happen if gene A was cut out and inserted the other way round against the promoter?

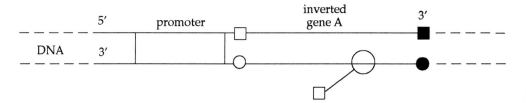

antisense mRNA

The answer is of course that RNA would be made using the wrong DNA strand as a template. The RNA that would be produced would be complementary (ie could hybridise) with the normal mRNA. Such an RNA molecule is said to be antisense RNA and the gene is said to be in the antisense orientation.

We can make use of the antisense orientation to help us to identify the relationship between a gene and a trait.

The theory behind this is that antisense mRNA will be able to form a complex with the wild type mRNA and can inhibit the expression of the trait. This results either in complete or partial inhibition of the expression of the trait, which can then be compared with the mutant alleles. This antisense method will be very useful for the analysis of plant genes that have been isolated by differential screening or chromosome walking. Introduction of antisense gene can also be used to modify the phenotype of a plant.

SAQ 7.9

Plant A is homozygous and produces white flowers. Plant B is homozygous and produces red flowers. When they were crossed, 100 F_1 generation plants were produced. Of these, 99 were red flowered, 1 was white flowered.

The DNA was extracted from red flowered and white flowered F_1 plants and treated with a restriction enzyme. The fragments were electrophoresised, blotted and hybridised with a radioactively labelled cloned transposon. Autoradiograms were produced of the blots and were as drawn below.

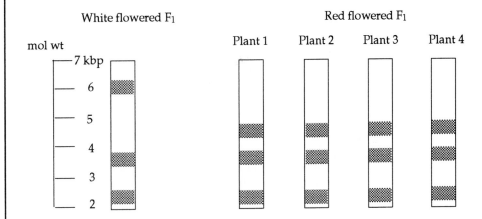

Explain how you could use this data to help to clone the gene which specifies red flowers.

Summary and objectives

This has been a long chapter but you have covered a lot of important issues.

Firstly you learnt how to use conventional cross breeding to assign genes to particular chromosomes and to determine how closely they were linked. Then you explored how Restriction Fragment Length Polymorphism analysis allowed mapping of genes even with genes for which the phenotypic expression was unknown. The bulk of the chapter however dealt with the isolation of individual genes either from cDNA or from genomic DNA. You learnt of the sequence beginning with mRNA right through to identifying the DNA species containing the gene of interest. This included both the cloning of the DNA in expression and non-expression vectors. We completed the chapter by describing the strategies for assigning particular genes to particular traits.

Now that you have completed this chapter you should be able to:

- explain how linked and unlinked genes segregate in the F_2 generation and to use data concerning F_2 progeny from two and three point crosses to determine gene linkages;

- interpret data from Southern blotting experiments in terms of gene segregation and restriction enzyme mapping;

- describe how a cDNA library can be produced and design oligonucleotide probes for cDNA identification from a knowledge of the amino acid sequence of the relevant gene product;

- identify putative organ specific genes from Northern blotting data;

- interpret the physical appearance of hybrid molecules derived from genomic DNA and cDNA molecules in terms of the presence and absence of introns and nucleotide sequence inversions;

- describe strategies for assigning genes to particular phenotypic traits;

- explain what is meant by antisense RNA and describe its potential in changing phenotypic expression of genes;

- use hybridisation data from transposon tagging to help to isolate a gene.

Gene transfer and genes to be transferred

8.1 Introduction   184

8.2 Gene transfer via *Agrobacterium tumefaciens*   186

8.3 Direct gene transfer   199

8.4 Genes for transfer   202

8.5 Other transferred traits   218

Summary and objectives   220

Gene transfer and genes to be transferred

8.1 Introduction

We have already learnt in earlier chapters that traditionally improved varieties of crop plants have been obtained by crossing parent plants each holding part of the desired traits and selecting the progeny of those individuals showing the proper combination of traits. Selection is possible because in progeny, traits segregate according to Mendelian rules. In this way tremendous improvements have been achieved in crop performance. Each day we can again experience the fruits of this breeding approach at the greengrocer or supermarket. Nevertheless plant breeding by crossing has its limitations. It is self evident that only traits present in related species can be combined, since crosses between distantly or unrelated species yield sterile plants or most often no progeny at all. Even somatic hybridisation has many limitations in terms of yielding viable plants. Another, often undesired, feature of cross-breeding is that all genes and thus traits, good and bad, of both partners are combined in the reassortment following fertilisation of ova by pollen. This inevitably will result in progeny plants having the desired traits combined with a number of unwanted ones. To eliminate these latter properties a time consuming programme of back-crossing is necessary. Because this strategy to get rid of undesired traits sometimes interferes with maximal performance of a crop, the time taken to introduce a new character into an existing variety by a back-cross programme often means that the variety is already out of date by the time the programme is completed. Thus, most often, progeny plants with a desired combination of traits show a number of undesired characteristics as well. We have illustrated this earlier by the Dutch potato cultivar Bintje isolated in 1917. It has superb qualities concerning yield, cooking and chipping and also for starch production and thus for consumer and industrial use this is the most widely grown potato in the Netherlands. Nevertheless Bintje is very susceptible to all kinds of plant pests like aphids, (the vector or viruses), fungi, bacteria and nematodes. This explains why the Netherlands Bintje cannot be grown without using substantial amounts of pesticides.

SAQ 8.1	From what you have learnt in earlier chapters, you should be able to make a list of the disadvantages of plant breeding by crossing and selection. Write down as many as you can.

In spite of the clear impressive achievements of traditional plant breeding by crossing and selection this approach has its limitations. Some of these limitations are removed by using molecular biological techniques. In the previous chapter we learnt how to map and isolate genes. In this chapter the molecular methods of gene transfer to crop plants will be described. Subsequently, examples of new traits introduced into crop plants by this new route will be described as well as traits which potentially will be introduced in the near future.

8.1.1 The potential and limitations of molecular plant breeding

Molecular plant breeding has opened up the possibility of:

- isolating and amplifying genes encoding for a particular trait in a test tube;

- equiping an isolated gene with specific regulation signals recognised by the transcriptional and translational machinery of the plant;

- introducing this modified gene into plant cells;

- regenerating these so called transgenic plant cells to mature and fertile plants.

advantages of molecular plant breeding

The main advantage of molecular plant breeding is that it offers the possibility of adding a new trait to an existing valuable variety. Also, in molecular plant breeding, species barriers do not exist, which means that in principle genes of any organism can be used for improvement of crop plants. Successful application of this approach has been well documented for dicotyledonous (broad leaf) crop plants like tobacco, potato, tomato and rapeseed which are quite amenable to these methods.

modification of promoter activities

Transfer of genes by these non sexual methods makes possible a whole new repertoire of plant breeding. Genes can be accessed from exotic sources - plants, animal, bacterial, viral or even human - and introduced into a crop. DNA elements that control gene expression, called promoters, can and often must be modified for proper function in the new host (bacterial genes do not function in plant cells unless adapted with plant recognition signals which are recognised by the plant transcriptional and translational apparatus). As we will see, it is possible to control timing, tissue specificity and expression level of transferred genes. Thus non sexual DNA transfer methods expand the sources of variability available for crop improvement. With methods available for chemically synthesising DNA, entirely novel genes can be created which potentially can lead to a whole new spectrum of designer traits in crop plants.

At this moment molecular plant breeding is hampered by three bottlenecks for a number of important crop species, notably the cereals. These are:

- no reliable protocols are available for the introduction of DNA;

- the number of isolated and agronomically interesting genes is still very limited;

- our knowledge of the regulation of gene expression in plants is limited;

- tissue culture methods for plant regeneration are still inadequate.

Introduction of a gene into a plant does not always have predictable results. Nevertheless, as we will see below, rapid progress has been made in this area.

| SAQ 8.2 | List the advantages of molecular plant breeding. What are the current limitations? |

8.2 Gene transfer via *Agrobacterium tumefaciens*

Methods of transfer of DNA and thus traits from one organism to another were already known in the 1940s. Transfer of genes or chromosome segments between bacteria by plasmids, sex factors, viruses (transducing phages) or uptake of purified DNA (transformation) were well understood before recombinant DNA technology was developed. In animal cell biology virally mediated transfer, direct DNA uptake and micro-injection of naked DNA have been successfully applied. In plants the most successful and advanced method for introducing foreign DNA is by a unique bacterial mediated DNA transfer system.

8.2.1 Natural gene transfer by *Agrobacterium tumefaciens*

For a long time it has been well known that infection of wounded dicotyledonous plants with the pathogenic soil bacterium *Agrobacterium tumefaciens* results in tumour formation, the so called crown gall (see Figure 8.1).

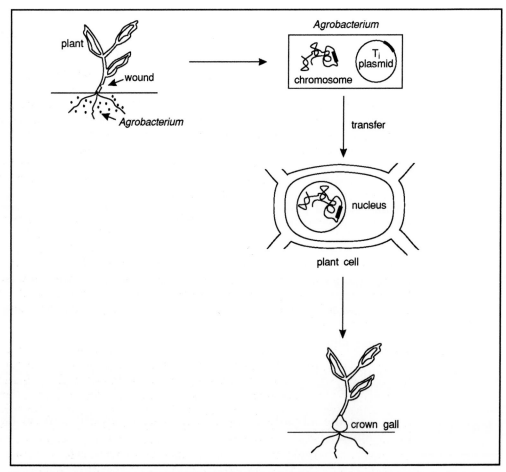

Figure 8.1 Transfer of part of *Agrobacterium* DNA to wounded plant cells and development of a tumour called crown gall.

opines

These tumours produce opines which normally are not found in plants but can be used by *Agrobacterium* as a carbon and nitrogen source. Opines are a group of amino acid derivatives. We will explore these in detail a little later. Thus the bacterium is capable of changing the plant into a production factory of compounds it can use. This change in the properties of cells in crown galls is caused by the introduction of a small piece of agrobacterial DNA called T(transfer)-DNA into the genomic DNA of the plant cell.

T-DNA

hairy roots

This piece of bacterial DNA contains the genetic information for opine production and also for tumour formation. The transfer of genetic information from a bacterium into a plant is highly unusual in nature. It has only been described for *Agrobacterium tumefaciens* and *Agrobacterium rhizogenes*. The latter is the causative agent of the so called hairy root disease (the tumour has the appearance of roots instead of a crown gall). Based on these agrobacterial properties, discovered in the late 1970s, very efficient protocols have been developed for the transformation of plants.

T_i and R_i plasmids

The mechanisms underlying crown gall and hairy root disease are similar. In both cases, plant cells at the infection site are transformed by a segment of plasmid DNA originating from the tumour inducing bacteria. Agrobacteria contain large plasmids (larger than 100 000 basepairs of DNA) on which genes involved in virulence are located. These plasmids have been named T_i (tumour inducing) plasmids for *Agrobacterium tumefaciens* while those present in *Agrobacterium rhizogenes* are called R_i (root inducing) plasmids.

oncogenic genes

hormone independent

ipt

iaaM

iaaH

Crown galls develop due to the expression in the plant cell of oncogenes (onc) that are present on the T-DNA, the part of the T_i plasmid which is transferred to the plant cell. The T-DNA segment (about 25 000 basepairs of DNA) contains the genes called ipt (for isopentenyl transferase), iaaM (for tryptophan mono-oxygenase) and iaaH (for indolacetamide hydrolase). These genes encode enzymes capable of producing the plant hormones isopentenyl AMP (a cytokinin) and indole acetic acid (an auxin). This explains why plant cells transformed with T-DNA can grow in the absence of these growth regulators in tissue culture. This contrasts with normal plant cells which are dependent on them for growth (see Figure 8.2).

rol genes

In the case of the R_i plasmid, the genes which are responsible for the transformation of normal cells to hairy roots have been identified and called rol (root-locus) genes. There are four of them called rolA, rolB, rolC and rolD. Until now it is not clear how these rol genes influence the plant cells to become hairy roots. There are indications that the products of the rol genes increase the sensitivity of the transformed cells to the plant hormone auxin. Thus, instead of production of growth hormone like the oncogenes present on the T_i plasmid, the rol genes modify the responsive mechanism of the cell to plant hormones.

plant regeneration from hairy roots

There also exists a second difference between crown galls and hairy roots. In contrast to crown galls, hairy roots can be regenerated to mature plants as can normal roots. The hairy root plants do show diminished apical dominance and an altered morphology compared to normal plants. The leaves are highly crinkled and curly, they grow a lot of roots which do not show geotropism and, in the case of potato, the tubers have deep lying eyes. Most often these plants are sterile. Nevertheless because hairy roots are formed quickly after infection (usually within two weeks) and because these roots can be regenerated to plants the *Agrobacterium rhizogenes* transfer system is a very useful system to study the expression of foreign genes in transgenic plants.

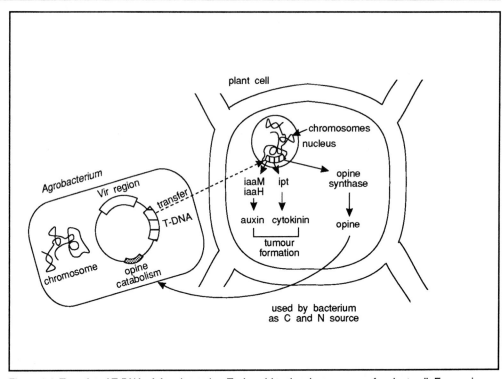

Figure 8.2 Transfer of T-DNA of *Agrobacterium* T$_i$ plasmid to the chromosome of a plant cell. Expression of the genes located on the T-DNA, iaaM and iaaH for auxin production and ipt for cytokinin production, results in tumour formation. Opine synthase activity makes opine synthesis possible. Opines can be used by the bacterium as carbon and nitrogen source (see text for further details).

SAQ 8.3

Briefly describe the molecular mechanism causing crown gall and hairy root disease.

metabolising
opine genes
carried by R$_i$
and T$_i$ plasmids

Besides the onc genes, the T-DNA also contains genes encoding enzymes involved in the production of tumour specific metabolites called opines which are either condensates of amino acids and a sugar (eg octopine, nopaline, leucinopine, agropine and mannopine) or phosphorylated sugar derivatives like agrocinopines (see Figure 8.3).

Π Examine Figure 8.3 carefully and identify the amino acids to which the various opines are derived. You should be able to identify arginine and leucine.

Opines are excreted from the transformed plant cells and can be used by *Agrobacterium* since opine catabolising enzymes are also present on the T$_i$ and R$_i$ plasmids. These two plasmids can be distinguished from each other by the opine catabolic/anabolic genes they carry (ie octopine, nopaline, and leucinopine metabolising enzymes are coded by T$_i$ plasmid genes. Agropine and mannopine metabolising genes are carried by R$_i$ plasmids).

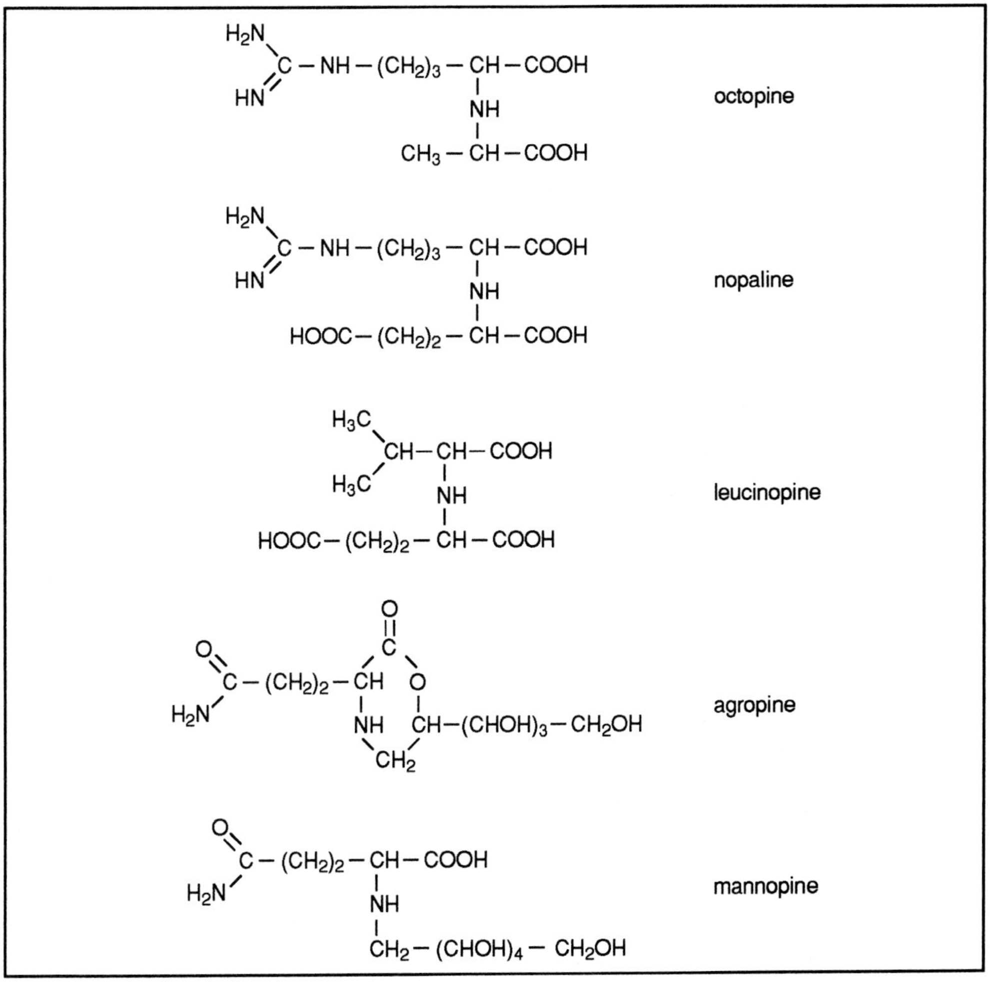

Figure 8.3 Chemical structure of different opines.

SAQ 8.4	In nature *Agrobacterium* transfers DNA into plants. Why? (Figure 8.2 will help you to answer this question).

8.2.2 Mechanisms of transfer of the bacterial T-DNA to plant cells

vir and chv genes

Besides the onc genes, a large number of other bacterial genes present on either the T_i plasmid or on the bacterial chromosome are involved in the induction of tumour formation. The vir (virulence) genes present on the T_i plasmid and the chv (chromosomal virulence) genes present on the bacterial chromosome are essential for the transfer of the T-DNA. (The roles of these genes are described below. Use Figure 8.4 to help you follow the description).

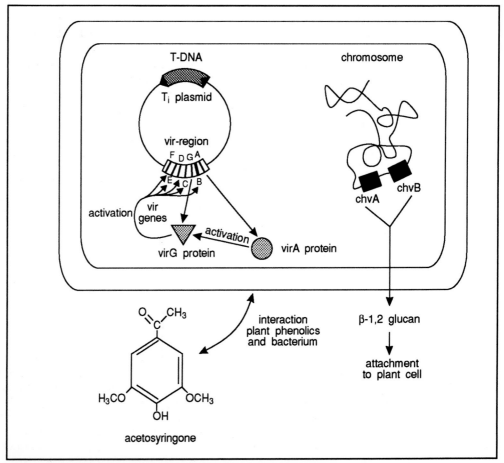

Figure 8.4 Induction of the vir region of the T$_i$ plasmid by acetosyringone via interaction with membrane located virA protein and signal transport virG protein (see text for details).

chv genes

cyclic glucan
transport
protein

For transfer of the T-DNA to plant cells, *Agrobacterium* first has to attach to plant cell walls. This is accomplished by the products of the chvA and chvB genes. The chvB gene encodes a protein involved in the formation of a cyclic β-1,2 glucan while the chvA determines a transport protein located in the bacterial inner membrane. This transport protein is necessary for the transport of the β-1,2 glucan into the periplasm (space between the cell wall and plasma membrane). Most likely the β-1,2 glucan plays an important role in the attachment of *Agrobacterium* to the plant cell wall. For transfer of the T-DNA this attachment is essential.

vir genes

many vir genes

Vir genes are involved in the actual transfer of the T-DNA from the bacterium to plant cells. For example, they appear to be essential for cutting the T-DNA out of the T_i plasmid, transfer over the bacterial membrane into the plant cell cytoplasm, transport to the nucleus and finally integration into the host genome. In total 22 vir genes on the octopine T_i plasmid are present in 7 operons. These are labelled virA to virG. These operons sometimes encode one, sometimes more, proteins (eg virB operon contains 11 open reading frames, while virA contains only one). Normally the vir genes are not expressed in *Agrobacterium* until they become activated by certain plant factors. These

acetosyringone

hydroxy-
acetosyringone

factors have been identified as phenolic compounds like acetosyringone and hydroxyacetosyringone which are released from plant tissue, especially after wounding. It has long been know that wounding is an absolute prerequisite for plant tumorigenesis via *Agrobacterium*. These phenolic compounds most likely act via two proteins called virA and virG encoded by 2 genes of the virulence region. The virA protein is located in the bacterial membrane and according to the present model can sense the presence of phenolic inducer compounds like acetosyringone. By binding the inducer the virA protein might become capable of phosphorylating the vir G protein which is present in the bacterial cytoplasm. In turn this modified virG protein can bind to the promoter regions of other vir genes inducing their expression. The role of the other vir gene products is less clear.

The virB operon consists of 11 open reading frames so putatively can encode 11 proteins. Most likely according to their predicted amino acid sequences, these proteins are exported to the membrane of the bacterial periplasm. These virB proteins are essential for T-DNA transfer and might therefore determine a physical structure (pilus or pore) that makes transfer possible. One of the virC proteins and two of the virD proteins are involved in the actual transfer of the T-DNA (see below). Besides these vir genes a number of accessory vir genes are present possibly functioning in the determination of the host range of the *Agrobacterium*. Chromosomal genes also play a role in this host range determination which until now is not well understood. In nature it is most likely that host range determination will have a function in choosing the best plant partner for production of opines.

8.2.3 T-DNA transfer (refer to Figure 8.5)

Besides T-DNA no other parts of the T_i or R_i plasmid become integrated into the genome of plant cells. Also no physical linkage of T-DNA and T_i plasmid is necessary for T-DNA transfer to occur. In fact none of the genes located on the T-DNA are involved in their own transfer, confirming that the transfer system exclusively is determined by the products of the vir and chv genes. However particular DNA sequences of 24 basepairs

border
sequences

which flank the T-region and are called the border sequences, are also essential for the transfer. They act as recognition and starting signals for the transfer apparatus. It is not clear at present, in what form the T-DNA is transferred to the plant cell. After induction of the vir genes by phenolic compounds, like acetosyringone, single stranded nicks are introduced into the strand of the double stranded border sequences of the T-DNA.

single stranded
T-DNA

Single stranded DNA molecules (called T-strands) representing one strand of the T-DNA can be found in *Agrobacterium*. Both nicking and single stranded T-strand DNA formation are dependent on the activity of two proteins called virD1 and virD2 encoded by the virD operon of the virulence region. These proteins together apparently determine an endonuclease activity capable of nicking the T-DNA border repeats at a precise site. This nick might then act as a starting point for the repair synthesis of the bottom strand and subsequently the release of the T-strand.

Whether the T-DNA introduced in the plant cell is single stranded or double stranded is not known, however there are strong indications that the *Agrobacterium* T-DNA transfer system strongly resembles conjugative DNA transfer between bacteria. In the latter case single stranded DNA is transferred from donor to recipient.

Most likely *Agrobacterium* does not introduce naked DNA molecules into plant cells. Evidence has accumulated recently that T-strands have the virD2 protein covalently attached to the 5' terminus. Furthermore the virE2 protein is a protein which binds single stranded DNA and is capable of coating the T-strands by co-operative binding leading to long thin nucleoprotein filaments. So presumably this T-DNA intermediate is protected by virE protein from degradation by DNA degrading enzymes present in the plant and bacterial cells.

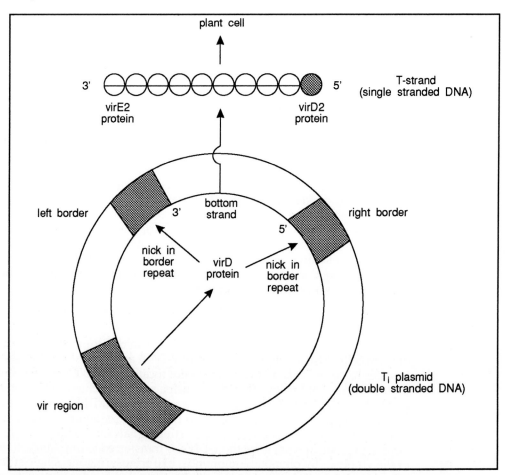

Figure 8.5 Production of T-strand by virD protein activity. The T-strand is generally believed to be the intermediate between bacteria and plant. Coating of the T-DNA by the virE protein protects it from degradation by DNases.

Π   This is quite a complex picture. To help you remember it, it would be advisable to make your own flow diagram. Begin with the vir region and, by re-reading the last two sections, write on your diagram the function of the genes. We have provided a start to such a flow diagram which you could complete.

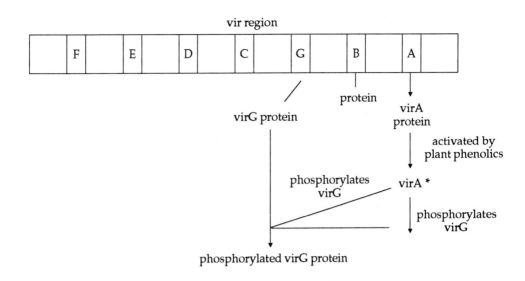

When you have completed this diagram, use it to help you to answer the two SAQs.

SAQ 8.5

Describe the roles of the bacterial vir genes in the transport of T-DNA into the plant cell.

SAQ 8.6

Describe the formation of T-strands and the role of endogenous DNA polymerase.

8.2.4 Host range of *Agrobacterium*

Agrobacterium infection is mainly confined to dicotyledonous

Dicotyledonous plants like tobacco, tomato, potato, petunia, rapeseed, cauliflower, lettuce, sunflower, chrysanthemum, rose, apple and many more are susceptible to *Agrobacterium*. However, most monocotyledonous plants like wheat, maize, rice and grasses do not respond with tumour formation upon infection with *Agrobacterium*. Up until now, transfer and integration of T-DNA in the plant genome has only been proved in the monocotyledons species asparagus and *Dioscorea*. Transgenic asparagus plants have now been isolated.

lack of callus formation in monocotyledons

Monocotyledonous plants possibly do not produce enough or the right vir gene inducers to allow *Agrobacterium* to transfer its T-DNA. More likely there exists a fundamental difference between di- and monocotyledons after wounding. In dicotyledons wounding leads to dedifferentiation of differentiated cells which heal the wound by production of a wound callus. Cells adjacent to the wound site can subsequently redifferentiate. If dedifferentiated (callus) cells are isolated from a wound, they can be used to regenerate new plants. It is this property which is exploited in plant tissue culture. Monocotyledons do not show the phenomenon of dedifferentiation after wounding and do not produce wound callus. Thus, tissue culture of monocotyledons usually requires the use of immature or meristematic tissues which have not lost the capacity for cell division. Thus, even if the T-DNA integrates, the transformed cell can not differentiate to a mature plant. This lack of susceptibility hampers the genetic manipulation of these important crop plants. Fortunately, alternative methods to introduce isolated genes into these plant cells have recently been developed; they will be discussed later.

8.2.5 *Agrobacterium* T$_i$ plasmid based vector systems to introduce foreign genes into plant cells (see Figure 8.6)

production of useful vectors

As we have seen above, besides the T-DNA no other parts of the T$_i$ or R$_i$ plasmids become integrated into the genome of the host cells. Actually, even physical linkage between T-DNA transfer and the rest of the T$_i$ plasmid is not necessary for T-DNA transfer to occur. Moreover as we have seen, T-DNA transfer does not require genes located on the T-DNA. The T-DNA transfer system is determined completely by the vir and chv genes. Only 24 basepair T-DNA border sequences are essential for this process.

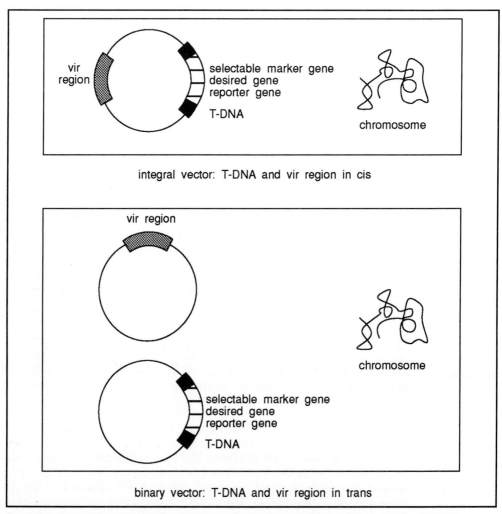

integral vector: T-DNA and vir region in cis

binary vector: T-DNA and vir region in trans

Figure 8.6 Composition of integrative and binary disarmed *Agrobacterium* vectors (see text for further details).

cis and trans systems

On the basis of these facts very useful vector systems for transferring foreign genes into plant cells have been developed (see Figure 8.6). These systems can be divided into two types:

- Cis systems in which the genes to be transferred are inserted into an artificial T-DNA which is part of a T_i plasmid. Thus physical linkage, which is called cis, exists between the modified T-DNA and the DNA encoding the transfer apparatus, both are part of one T_i plasmid molecule;

- Trans or binary systems in which foreign genes are cloned in between border sequences into plasmids, which are capable of replication in *E.coli* and *Agrobacterium*. Thus manipulation can take place in the easy-to-handle *E.coli* bacterium and subsequently the plasmid containing the genes to be transferred can be conjugated to *Agrobacterium*. This *Agrobacterium* strain harbours a second plasmid containing the vir genes. Thus the foreign genes to be transferred and the genes involved in their transformation are not physically linked but located on two different plasmids, so called located 'in trans'. T_i plasmids used in the transfer of foreign genes into plant cells are called disarmed because tumour formation is not induced because the oncogenic T-DNA is removed. Both of these systems have successfully been used for the introduction of genes into plants.

disarmed vectors

artificial T-DNA

The artificial T-DNA used in these vector systems minimally consists of the 24 basepair border repeats. All genes involved in tumour formation and opine synthesis that normally are present on the T-DNA are removed. These genes are part of tumour inducing and opine synthesising machinery and are not needed for transfer. Moreover they have to be removed for recovering normal looking plants after transfer. Thus plant cells transformed by this artificial T-DNA behave as normal plant cells in tissue culture and can be regenerated to normal looking plants after transfer.

kanamycin resistance gene transferred

In the early 1980s the first transfer of a bacterial gene to a plant cell using disarmed (no tumour formation, no opine synthesis) T-DNA vectors was reported. The bacterial gene used encoded the enzyme neomycin phosphotransferase (NPT) which is able to phosphorylate and thus inactivate the antibiotic kanamycin. The rationale for use of this gene was the expectation that expression of this gene in plant cells would render the cells resistant to the antibiotic. Normally plant cells are very sensitive to it and killed if kanamycin is added to the synthetic medium on which they are grown. Those cells transformed by manipulated strains of *Agrobacterium* and thus containing and expressing the NPT gene became resistant towards kanamycin and could thus be easily selected out of a population of untransformed and transformed cells. It has to be stressed that only a very minor population of all plant cells which are adjacent to the wound site and being in contact with *Agrobacterium* are actually transformed. So an ability to select the transformed cells from a huge background of untransformed cells is a prerequisite.

requirements to be able to select transformed cells

8.2.6 Expression of marker and reporter genes in plant cells

In contrast to the agrobacterial genes located on the T-DNA and encoding tumour formation and opine synthesis in the first instance the bacterial kanamycin resistance gene was not functional in plant cells. This holds for all genes normally not present in plants. Detailed analysis showed that they were not transcribed by the plant RNA polymerase. A simple explanation for this discrepancy was that foreign genes, like the bacterial kanamycin resistance gene do not possess the correct recognition signals (promoters and terminators) necessary for proper expression in plants. Thus chimeric genes were constructed to test this idea. These artificial genes originally consisted of the promoter and terminator of the opine synthesis genes derived from the T_i plasmid T-DNA region. These genes are obviously recognised by the gene expression apparatus of the plant. The NPT gene was used as protein coding region in these chimeric genes.

requirement for recognition signals

use of chimeric gene

These artificial genes turned out to be active in plants and caused resistance towards kanamycin.

SAQ 8.7
Is a flower specific promoter useful in driving a selection marker gene? (Give reasons for your answer).

In conclusion bacterial antibiotic resistance genes and other foreign genes can be expressed in plants if they are adapted with regulatory sequences recognised by the transcriptional and translational machinery of the cell.

selectable marker genes

Antibiotic resistance genes can be used to select transformed plant cells and thus are called selectable marker genes. Not only kanamycin resistance genes have been used to select transformed cells and to optimise the transformation procedure. The bacterial gene encoding chloramphenicol acetyltransferase (CAT) has also been used for this purpose. Plant cells are sensitive to chloramphenicol but they are not as sensitive as they are to kanamycin. In addition recently new selectable marker genes have been developed. Examples are the hygromycin phosphotransferase (HPT) gene which renders plant cells resistant to the antibiotic hygromycin and the phosphinothricine acetyltransferase gene (PAT) causing resistance to the herbicide phosphinothricine.

reporter gene

The CAT gene is not used anymore as a selectable marker gene. Nevertheless it is a very useful gene because its activity can be detected easily. Normally plant cells do not show CAT activity and thus the CAT gene can be used as a reporter gene of successful transformations (ie reporting the incorporation of foreign DNA into a plant cell). The same is true for the *E. coli* β-glucuronidase gene (GUS). Plant cells do not show endogenous GUS activity. GUS activity can be easily detected by very sensitive methods like fluorometry demanding only very limited amounts of tissue. GUS activity can also be detected by histochemistry in intact tissue and thus individual cells expressing GUS activity can be identified. At the moment the GUS gene is the most popular reporter gene in plant molecular biology.

Other useful reporter genes are the LUX genes of the firefly and of *Vibrio harveyi* encoding the light emitting luciferase enzymes. Like GUS activity, the light emission, caused by this enzyme can be detected very easily.

Thus in contrast to the NPT, HPT and PAT genes which can be used to select transformed cells and thus are called selectable marker genes, the CAT, GUS and LUX genes can only be used for reporting the transformation of plant cells or the activity of a promoter which drives the expression of it and thus are called reporter genes.

SAQ 8.8
Is a flower specific promoter useful in driving a reporter gene?

8.2.7 Programmed expression of foreign genes

importance and range of promoters

Fundamental research in animal and plant molecular biology has shown that gene expression in most cases is regulated by DNA sequences 5′ upstream of the coding region of a particular gene. This region is called the promoter and determines the expression of a gene. This expression can be in all cells of the plant but can also only be in a certain subset, for instance only in leaves or roots or tubers or flowers or seed. Promoters also determine genes to be expressed in light or dark or both, or to be specifically induced after heat, frost, wounding, pathogen attack etc. Isolation of promoters and analysis of their properties makes it possible to express foreign genes only in certain tissue, under certain conditions or during certain developmental stages (flower, seed, pollen, tuber). This is very attractive for molecular breeders because proteins toxic to plant pathogens can for example, be expressed in leaves only and not in edible plant parts like fruits and tubers. Therefore, a lot of effort has been put into the

35S and 2′ promoters

isolation of DNA sequences involved in the regulation of gene expression. In the first instance people concentrated on promoters which gave higher expression levels compared to the nopaline synthase gene promoter which was used in the first instance. Examples are the promoter of the cauliflower mosaic virus 35S RNA gene (35S promoter) and the 2′-promoter derived from the 2′ gene of the T-DNA of the T_i plasmid.

Both these promoters show enhanced transcription levels compared to the nopaline synthase gene promoter and thus, if they are used to drive kanomycin resistance genes, confer higher levels of kanamycin resistance to transformed plant cells making their selection easier.

proteinase inhibitor II gene promoter

The 35S promoter is expressed in all tissues of a transgenic plant but not in all cell types of all tissues. The 2′ promoter is normally expressed in roots but, upon wounding of plants, it is also expressed in cells adjacent to the wound site. Therefore, this promoter is ideally suited to drive the expression of genes encoding proteins toxic for plant pathogens. Another promoter which is inducible by wounding is derived from the proteinase inhibitor II gene of potato. This gene is silent in leaves but is systematically induced throughout the plant as the result of wounding of a single leaf.

heat shock promoters

Other genes are induced by a heat shock. The protein products of these kinds of genes are thought to be involved in protection of plants and animals against damage by high temperature. Isolated promoters of these genes are able to induce, upon heat shock, within 1 hour, the expression of reporter genes.

light driven promoters

A large number of genes expressed in leaves are induced by light. Promoter sequences of these genes (examples are the nuclear gene coding for the small subunit of the ribulose 1,5-bisphosphate carboxylase or the gene for the light harvesting chlorophyll a/b protein) which confer their light regulated expression are contained within small DNA sequences of no more than a few hundred basepairs. Some of them were shown to work as light dependent enhancers of gene expression in the leaves and as silencers of gene expression in roots (darkness). Chalcone synthase represents another type of light regulated gene. Using 5′ upstream sequences from the coding region it has been possible to construct chimeric genes that would be silent in transgenic plants unless the plants were irradiated for 20 hours by ultraviolet light.

organ specific promoters

Strict organ specific regulation of gene expression has also been shown for a number of genes. The patatin gene is solely expressed in potato tubers and the phaseolin gene only in seeds of *Phaseolus vulgaris*. Chimeric genes consisting of the promoters of these genes show strict organ specific expression of reporter genes.

Recently genes have been isolated which are only expressed in certain parts of the flower, making flower specific expression of foreign genes feasible.

Thus by isolating the regulatory DNA sequences responsible for restricted gene expression it is possible to direct a foreign protein to a predefined organ of the plant or to express a foreign gene only under certain environmental conditions.

∏ It should be obvious that the isolation of promoters and their use to 'drive' foreign genes inserted into plants, is of great importance to molecular plant breeding. To help you remember the range that are available, make yourself a table summarising this section. We suggest you use the following format:

Controlling factor	Example of a promoter
Wounding	proteinase inhibitor II gene from potato
Light	small subunit of ribulose 1,5-bisphosphate carboxylase

8.2.8 Targeting gene products to the chloroplast and mitochondria

Besides nucleus and cytoplasm, plant cells also contain chloroplasts and mitochondria. These organelles contain their own DNA which encode proteins necessary for their proper functioning (eg harvesting light and generating energy). However, for these processes to occur proteins encoded by the nuclear genome are also necessary. An example is the earlier mentioned small subunit polypeptide of ribulose 1,5-bisphosphate carboxylase which is encoded by a nuclear gene. Together with the large subunit encoded by the chloroplast, it forms the enzyme rubisco (ribulose 1,5-bisphosphate carboxylase) responsible for CO_2 fixation in plants. The small subunit polypeptide is formed in the cytoplasm. An obvious question is how this polypeptide is transported from cytoplasm to chloroplast and also: is it possible to use the transportation signals for the targeting of foreign proteins to the chloroplast and mitochondrion?

transit peptides

Proteins which are encoded by the nucleus and having as destination the chloroplast or mitochondrion are made as precursors containing an extra amino acid chain at their N-terminal end. This extra peptide, called transit peptide, is essential for transport across the organellar membrane into chloroplast or mitochondrion. The transit peptide is cleaved off from the precursor during or shortly after translocation. By fusing the transcribed coding sequence for this transit peptide to a reporter gene product it has been shown that the transcribed foreign protein could be targeted to chloroplast or mitochondrion depending on the peptide used (derived from a chloroplast or mitochondrial nuclear encoded protein).

8.2.9 Transformation methods using *Agrobacterium*

In nature *Agrobacterium* transfers its DNA to plants by infecting wounded cells. Wounding is an absolute requirement. Based on this property very efficient protocols have been designed to transfer isolated genes incorporated in disarmed T-DNA to the nuclear genome of plant cells. These protocols are based on knowledge about the possibility of regenerating certain plant tissues to plants. It was shown that non sexual,

totipotent plant cells

somatic plant cells are totipotent which means that in principle they can differentiate and form a whole new fertile plant. Thus, pieces of leaf tissue, when incubated under

suitability of leaf tissues

proper conditions (eg in the presence of plant growth regulators like auxin and cytokinin) can regenerate to shoots which can be rooted and grown to mature, fertile plants. Pieces of leaf tissue are in principle wounded due to the cutting procedure and thus ideally suited to be used for transformation. In the early 1980s it was shown that simple incubation of these leaf pieces with *Agrobacterium* resulted in crown gall development at the wound site and therefore, transfer of T-DNA had happened. Subsequently it was shown that disarmed T-DNAs containing the NPT gene were transferred to and by using the proper regeneration conditions in the presence of kanamycin, antibiotic resistant plants could be recovered. In the presence of kanamycin, only those cells survived which contained and expressed the NPT gene. In this way an easy and efficient protocol for plant transformation was set up. Note that after 'co-cultivation' with *Agrobacterium* to induce transformation, the bacterial cells must be killed with a different antibiotic in order to culture the transformed cells aseptically.

The next step was the transfer of an agronomically interesting gene, for instance an insect resistance gene, linked to the NPT gene. It was shown that if both genes were part of the same disarmed T-DNA transfer of both genes occurred. Thus, selection for kanamycin resistance could be used to detect transfer of the agronomically interesting gene.

The combination of the development of efficient regeneration protocols of wounded plant tissue and of disarmed *Agrobacterium* vectors containing a selectable marker gene made possible the transfer of foreign genes to plants.

At present all kinds of plant tissue responsive to regeneration are used for transformation and regeneration. Examples are leaf discs, stem pieces, potato tuber slices, root segments and flower tissue. A large number of plants have been transformed including tobacco, tomato, potato, rapeseed, sunflower, lettuce, beet, poplar, flax and cotton.

8.2.10 Inheritance of transferred genes

Transferred genes become stably integrated into the genome of the plant host and thus in general they behave like normal genes. The progeny of transgenic plants normally inherit the T-DNA according to Mendelian rules. This implies that selfed progeny show 3:1 ratios for a dominant trait like kanamycin resistance, while back crossed progeny show a 1:1 segregation. This is also true for multiple T-DNAs that are all genetically linked.

transferred
genes inherited
like
endogenous
genes

After transformation there can also be multiple T-DNAs, some of which are incorporated at different loci. In these cases selfed ratios are usually 15:1 (two unlinked loci) or higher (three or more unlinked loci). In nearly all cases the selected and unselected markers on a single T-DNA are co-inherited. This shows that T-DNA behaves genetically like endogenous genes. There are however cases where the two genes become dissociated, and there is increasing evidence for mitotic crossing over or segregation and even for meiotic events occurring *in vitro*.

8.3 Direct gene transfer

Monocotyledonous plants like cereals are not susceptible to the transformation techniques described above. Fortunately new methods to introduce DNA into plant cells have recently been developed which do not show this limitation. These methods will now be discussed.

no host
specificity

Purified DNA and thus genes can be used directly for plant transformation by, for example, direct DNA uptake involving physico-chemical reactions that result in DNA transfer to protoplasts (individual plant cells lacking a cell wall) or by injection with microscopic pipettes directly into cells. Unlike methods based on the *Agrobacterium* transfer system, direct gene transfer methods are not subject to host range restrictions, but practically are limited by the need to recover a whole plant from target protoplasts or tissue.

8.3.1 Gene transfer by PEG-directed DNA uptake and electroporation

PEG the
chemical of
choice

In Chapter 6, we discussed the use of polyethylene glycol in fusing protoplasts. Similar treatments can cause protoplasts to take up DNA. In the late 1970s it was shown that polycations like poly-L-lysine and poly-L-ornithine enhanced the uptake of DNA by plant protoplasts. However the transformation frequency only reached a maximum of about 1 in every 100 000 protoplasts. A similar transformation frequency was reported for polyethylene glycol (PEG) in a study which compared different uptake methods. Nevertheless PEG turned out to be the chemical of choice.

Compared to the *Agrobacterium tumefaciens* transformation method, the PEG mediated protoplast DNA transfer protocol resulted in a very low transformation efficiency. Because protoplasts are not always easy to use, this transfer system is seldom used to transform dicotyledonous plants.

However, the impossibility of transferring genes to cereals by *Agrobacterium* has led to a continuous interest in direct DNA uptake methods for the transformation of this important group of crop plants.

electroporation

Besides PEG alternative procedures have also been developed for the transfer of genes to protoplasts. Electroporation is one of them. This method subjects protoplasts to a high voltage electrical pulse to induce a reversible permeability change in the cell membrane.

Stable transformation by the direct DNA uptake method has now been shown among others for wheat, maize and rice protoplasts. However the regeneration of transgenic maize and rice protoplasts to mature fertile plants has only very recently (1989) been reported.

Thus DNA can be introduced into monocotyledonous protoplasts and incorporated in the nuclear DNA. However development of successful and reproducible regeneration methods for these transgenic protoplasts is still not routine. Moreover protoplast regeneration protocols for a large number of monocotyledons are still not available.

SAQ 8.9

What is the fate of naked DNA after PEG induced or electroporation guided uptake by protoplasts? (Think about where the DNA may end up - are there DNases present)?

8.3.2 Micro- and macro-injection of DNA

micro-injection

Micro-injection involves the introduction of plasmid DNA into plant protoplasts by means of micropipettes. In one case, cell lines from micro-injected tobacco protoplasts were shown to contain foreign DNA sequences stably integrated into the nuclear DNA. Fairly high transformation efficiencies up to 20% can be achieved. However to date there have been few reports of transformed plants regenerated from cell lines obtained by protoplast micro-injection. Micro-injection has also been applied to the transfer of DNA to 12-celled microspores. Up to 50% of the injected microspores gave rise to transgenic plants. These plants showed a chimeric character (ie not all of the cells of each plant contained the same genetic information). This was due to the fact that not all individual cells of the microspore had been injected and incorporated DNA in their nucleus. Individual cells of the chimeric primary somatic embryos could be induced to form intact secondary embryos. This would ultimately lead to completely transformed embryos which could be regenerated to plants. In spite of the fact that these results are promising no plants other than rapeseed have been transformed using this procedure.

| SAQ 8.10 | Is micro-injection the method of choice to transform rapeseed? |

macro-injection In an attempt to avoid the need to develop a protoplast to plant system for cereals other than maize and rice, transgenic rye plants have been obtained by injecting plasmid DNA into young floral tillers 14 days after meiosis. About 5% of the injected plants delivered seedlings showing integrated plasmid DNA. This method has been called macro-injection because cells that were part of intact plants were used. It has to be stressed that this method has been reported by one research group only and most likely will turn out not to be applicable in general.

8.3.3 Particle gun for DNA delivery

The particle gun is the most recent addition to the repertoire of successful plant transformation methods. The procedure used for delivery of DNA to intact plant cells makes use of high velocity microprojectiles, thereby avoiding the requirement to isolate protoplasts for transformation. DNA coated tungsten particles (4µm in diameter) have been used to carry DNA into epidermal tissue of onion. Bombardment of shoot meristems of immature soyabean embryos by DNA coated gold particles and subsequent regeneration of these embryos have resulted in transformed shoots at a frequency of 2%. Ultimately only one fertile transformed soyabean plant has been recovered. Transgenic tobacco plants have also been obtained after plant regeneration from cells of leaves and suspension cultures treated with tungsten microprojectiles. In spite of these scarce successes, a large number of laboratories are applying this approach to achieve transformation of monocotyledonous plants.

ballistic A number of reasons can be put forward to be sceptical about the ballistic approach. First it may be difficult to deliver enough DNA on a particle to reach a high enough concentration of the gene to get a sufficient frequency of integration, also the number of regeneratable cells in cereals is extremely low.

abraded embryos Nevertheless we should not be too pessimistic, much is happening. Recent reports using abraded embryos (in which the cell layer which ultimately gives rise to the germ line is exposed) to provide material for transformation indicate that a break through in molecular breeding in cereals may be imminent. In these studies, transformation was measured by the expression of GUS reporter genes.

8.3.4 Other transformation approaches

Several other methods for transforming cereals have been reported. None of them have been confirmed by more than one research group. Nevertheless they will be described here to show the different approaches.

pollen transformation One of them is pollen transformation. In spite of a large number of experiments in different laboratories, no single transgenic offspring have been produced. This indicates that this approach has a low chance of success.

pollen tube

Another approach uses the pollen tube pathway. Access to the female egg cells in the embryo sac occurs via the pollen tube which grows from the pollen down through the pistil. Could these tubes be used for DNA delivery to the zygote? The presence of several biological obstacles (eg callose plugs within the pollen tube, nucleases, adsorption by cell walls) suggest that this approach is not a very promising one. Currently no convincing transgenic offspring has been presented in the scientific literature.

incubation of seeds with DNA

Incubation of seeds with a DNA solution was also tried for a long time. It is not easy to imagine how DNA uptake would happen through the rigid seed wall. Recently another approach has been described using dry cereal embryos separated from the kernel at the scutellum, thus creating a large wound site. So far there is no proof of integrative transformation and the chance of producing transgenic cereals would appear extremely small because again the DNA has to transform those cells involved in shoot regeneration and gametophyte formation.

We conclude that direct gene transfer can be very successful if protoplasts are used and regeneration protocols are available. Transgenic maize and rice have been recovered by this method. Because protoplast regeneration is possible for only a limited number of maize and rice varieties, it is absolutely no routine procedure. To develop such methods will be the challenge of the future. Perhaps micro-injection and/or particle bombardment of developing embryos will be the best methods to use for these target crops.

8.4 Genes for transfer

monogenic and polygenic traits

As discussed in the introduction to this chapter, one of the prerequisites for molecular plant breeding is the availability of genes which are able to confer desirable traits. Many phenotypic traits result from the expression of more than one gene and from the interaction of the corresponding gene products. For instance, several gene products may contribute to the formation of one functional enzyme consisting of different subunits. Also, in many biochemical pathways several enzymes are involved. As a consequence, in many cases the transfer of more than one gene will be necessary to obtain a desired trait.

| SAQ 8.11 |

Explain why several genes are needed to confer an entire new biosynthetic pathway.

examples of monogenic traits

The development of molecular plant breeding is still in its infancy. Currently most vector systems do not allow the simultaneous transfer of more than one gene or a few genes. In only rare cases have all genes involved in a particular biochemical pathway been cloned, and fine tuning of the regulation of the expression of these genes has even not yet been undertaken. This explains why molecular plant breeding focuses on the transfer of so called monogenic traits, ie phenotypic traits which result from the expression of a single gene. Several resistances can be introduced as monogenic traits, and therefore these have been the subject of much research in molecular plant breeding. In the following paragraphs results of molecular resistance breeding will be discussed. Transgenic plants resistant to viruses, insects or herbicides have been obtained and likewise plants resistant to bacterial pathogens are presently being produced. In addition, attention will be paid to a potential source of additional resistance genes, coding for the so called pathogenesis related proteins. Finally some examples of successful molecular breeding outside the field of resistance breeding will be given.

These examples concern transgenic plants showing altered flower colour (petunia), altered starch composition (potato), or postponed fruit softening (tomato).

8.4.1 Virus resistance

Coat protein protection

Each virus contains at least the nucleotide sequence (either DNA or RNA) coding for genes essential for its multiplication and proteins surrounding the viral nucleic acid. These proteins are called coat proteins. As a first step in the infection process, the coat proteins are removed after entering the cell. The coat of a large number of plant viruses contains only a single protein species, ie only one gene codes for the identical subunits of the viral coat.

strategies for providing virus resistant plants

cross protection

To achieve the goal of virus resistant transgenic plants, three different strategies have been developed, of which the strategy making use of 'coat protein protection' has been most successful. This strategy is based on the finding that infection of a plant by a mild (attenuated) virus strain can suppress or postpone symptoms caused by superinfection with a second more virulent strain. The molecular basis for this finding is not yet understood. Nevertheless, the finding called cross protection, has been used in practice in tomato and citrus to diminish viral damage. However, large scale application has been hampered by the fact that the virus used for cross protection can only be allowed to cause slight symptoms and at the same time must be sufficiently related to induce cross protection. Besides, a virus showing mild symptoms in one plant may cause severe disease in another species, it may undergo mutations that render it more virulent, or it may act synergistically with a non related virus resulting in more damaging effects. Some of these disadvantages can be overcome if only one viral gene is transferred to the plant instead of an intact virus.

SAQ 8.12

Which disadvantages associated with using attenuated viruses can be overcome by using a single viral gene? Which disadvantage can not be overcome by this approach? Indicate why.

Genetic manipulation has opened up this route. For several plant viruses combinations of a T-DNA construct containing a DNA copy coding for the viral coat protein have been introduced into the host plant. The resulting transgenic plants contained one or more integrated copies in the genome and transcripts of the viral coat protein 'gene' were demonstrated. Since a constitutive promoter (35S CaMV) was used to control the coat protein expression, the resulting plants produced the viral coat protein constitutively (ie in all developmental stages and in most organs). As a consequence of viral coat protein expression, plants can become resistant to an infection with the same or a closely related virus. Usually, in resistant plants 0.1 to 0.5% of total cell protein is shown to be viral coat protein. 'Coat protein protection' has been proven effective for the combinations tobacco mosaic virus (TMV)/tobacco, TMV/tomato, alfalfa mosaic virus (AlMV)/tobacco, tobacco spotted virus (TSV)/tobacco, cucumber mosaic virus (CMV)/tobacco and potato virus X (PVX)/potato.

constitutive expression

One of the lines (6796) produces an antisense mRNA. Do you remember what an antisense RNA is? (We described this in Chapter 7). To remind you, it is an RNA strand which contains a complementary nucleotide sequence to a normal mRNA. Therefore the antisense RNA will hybridise with the normal mRNA preventing its translation into a protein sequence.

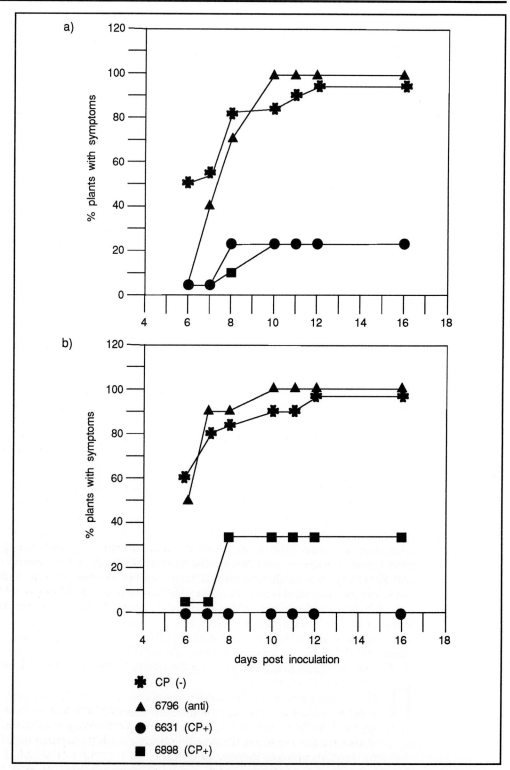

Figure 8.7 Protection from viral infection in transgenic plants. Control and transgenic plants were exposed to cucumber mosaic viruses (CMV). Line CP⁻ does not express viral coat protein. Lines 6631 and 6898 (CP⁺) express viral coat protein. Line 6796 (anti) produces an antisense coat protein mRNA. a) plants were inoculated with 5µg/ml CMV, b) plants were inoculated with 25µg/ml CMV (after Cuozzo *et al* 1988 Biotechnology 6, 549-557).

Π From Figure 8.7 are all viral coat protein producing transgenic plants equally resistant to viral infection? Is using the coat protein more efficient at providing protection than producing the antisense mRNA?

The answer to the first part is no. If you follow Figure 8.7 carefully you will see that the coat protein producing lines (6631 and 6898) show some differences in the % of plants which develop symptoms after exposure. The answer to the second part is also no. The line producing antisense coat protein mRNA shows a greater tendency to succumb to infection than does either line 6631 or 6898.

Although the mechanism of this protection has not been fully understood, the following conclusions have been drawn from the kinds of experiments described by Figure 8.7:

- 'coat protein protection' is exhibited at the protein level. If no translation of coat protein mRNA takes place, protection is not accomplished;

- 'coat protein protection' is a true form of resistance and does not lead to a form of tolerance in which replication coincides with the absence of (severe) symptoms;

- the protection is virus specific. Unrelated viruses show normal infection patterns;

- when viral RNA instead of intact virus is used as inoculum, protection does not take place.

A possible mechanism has been postulated at which an intracellular 'receptor' for incoming virus particles might be saturated with coat protein, thereby blocking a normal infection process. This is illustrated in Figure 8.8. The 'receptors' normally recognise and attach to the coat proteins of the virus. In transgenic plants, the plant cell produces coat proteins which fill the receptors, thus preventing viruses from binding.

Satellite protection

Several plant viruses can be accompanied by 'satellites', that are able to influence the symptoms of the viral infection. Satellites are small RNA molecules (of a few hundred bases in size) that co-replicate with the accompanying 'helper' virus. In doing so, they suppress the replication of the helper virus, probably by competing for the virus encoded replicase, the enzyme that copies the viral RNA molecules. This may subsequently lead to attenuation of the symptoms of a viral infection. Exceptions to this general rule have been observed in which replication is not diminished but attenuation nevertheless takes place. There is no explanation as yet for the mechanism of attenuation in the latter cases.

Π It might be useful to construct a diagram summarising the following paragraph to help you remember it.

Some efficient satellites which dramatically reduce symptoms of the viral infection they accompany, have been used as templates for the generation of tandem DNA copies that have subsequently been transferred to plants. Tandem copies are DNA molecules having multiple copies of the relevant nucleotide sequence. Transgenic tobacco plants containing these tandem copies and showing resistance to CMV or tobacco ringspot virus (TobRV) have thus been engineered. Whereas in non infected transgenics only a low level of the tandem satellite transcript could be observed, plants infected with the

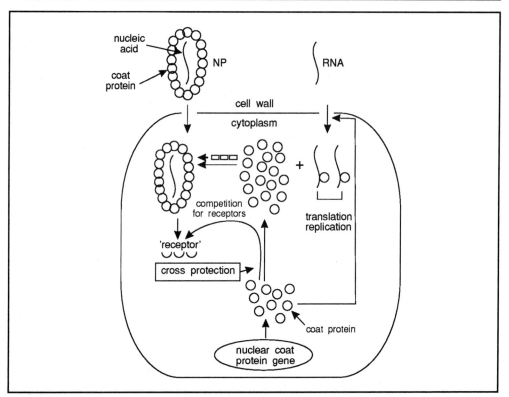

Figure 8.8 Model of coat protein protection. Note that if the nuclear encoded coat protein prevents more than one type of virus for binding with its receptor, the coat protein will provide protection against more than one infection.

helper virus without satellites showed high levels of monomeric satellite copies whose emergence coincided with attenuation and hence were apparently biologically active. When a unique copy of the satellite molecule was introduced into the plant genome, no monomeric transcripts were produced in the transgenic plant cell. A virus encoded function, dependent on a tandem template array as observed during normal satellite replication, apparently evoked the production of the monomeric satellite molecules, resulting in suppressed replication of the virus.

The transgenic cell thus mimics the natural situation in which satellites coinfect with the virus. Satellites are not known for all plant viruses. Artificial satellite RNAs derived from parts of a viral sequence might be developed for those viruses that have no natural satellites.

SAQ 8.13

Are the intracellular receptors illustrated in the model described in Figure 8.8 virus specific? Indicate why.

Other approaches

antisense strategy not successful against viruses

An approach that has shown little success in the search for viral control is the antisense strategy. Some years ago it was shown that the expression of a given gene can be blocked by the introduction into the cell of a complementary sequence, producing an antisense RNA. Although success in other applications has been reported (see below), transgenic tobacco plants containing genes coding for antisense viral RNA of TMV, tobacco rattle virus (TRV), or CMV are only weakly resistant or even fully sensitive to infection. The level of antisense RNA expression obtained might not be high enough to provide complete blockage of viral replication. However, the mechanism of antisense action is still poorly understood, and promising results of this strategy in other areas have lead to continuing research in the quest for viral control.

Finally, natural plant resistance genes, some of which confer monogenic dominant traits will be the subject of growing research. Already about 30 of such monogenic resistances have been identified, but no natural virus resistance genes have yet been isolated. Lack of knowledge about the mechanism of action and difficulties in identifying the gene products involved still hampers progress along this promising line.

SAQ 8.14	1) Are protein products synthesised from antisense mRNA?
	2) Do antisense constructs include a promoter region?

8.4.2 Insect resistance
(Insect resistance based on the expression of *Bacillus thuringiensis* crystal protein genes)

entomo-pathogenic bacteria advantages over chemical pesticides

The most well known entomopathogenic (pathogenic to insects) bacterial species is the sporeformer *Bacillus thuringiensis*. It was first isolated in 1901 from Japanese silk worm cultures showing unusually high mortality rates. Since then thousands of isolates have been found and some have been used for the production of bio-insecticides. There advantages over chemicals are their narrow host range (no parasites or predators are affected), the absence of development of resistance against *B. thuringiensis* in insect field populations, and their ecological safety both during production and in handling.

SAQ 8.15	Under which conditions is a narrow host range a disadvantage?

effects of B. thuringiensis proteins on insects

The toxic principle of *B. thuringiensis* is located in a large, intracellular protein crystal, which is formed during spore development. Depending on the isolate, these crystals contain several different, often homogous proteins. In crystals of isolates active against lepidopteran insect larvae most of these proteins have a molecular weight of 130 000 daltons (approximately). When ingested by a larva, the protein crystal is solubilised by the high pH of the larval intestinal juices, and insect proteases subsequently convert the liberated protein monomers of MW 130 000 daltons, which are called protoxins, into the true toxins of approximately MW 65 000 daltons. In sensitive larvae these toxins recognise a specific set of membrane receptor proteins present on the epithelial cells of the insect midgut. After binding to these receptor proteins the cells of the midgut then swell and eventually collapse, followed by disintegration of the intestinal tract and death of the larva. Figure 8.9 shows the structural relationship of different *B. thuringiensis* proteins.

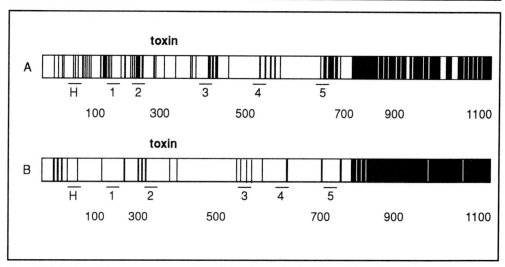

Figure 8.9 Amino acid sequence comparison of *Bacillus thuringiensis* crystal proteins. The open bars represents the complete amino acid sequence of the crystal proteins. Vertical lines within these bars represent amino acids that are conserved for all lepidopteran crystal proteins (A) and all dipteran crystal proteins (B). The toxicity of the protein is located in the left N-terminal part of the crystal protein. The position of five conserved sequence blocks (1-5) in both types of crystal proteins are underlined. The numbers (100, 200, etc) correspond to the amino acid number. (After Höfte *et al* (1989) Microbial Reviews 255-272).

The effectiveness of a *B. thuringiensis* crystal protein is thus dependent on its solubilisation, on the pH and the proteinases of the larva and on the set of membrane receptors present on the midgut epithelial cells. In other words, these factors determine the insect specificity of a given crystal protein. Depending on the isolate, one to seven different crystal proteins can be demonstrated in a single crystal, which presumably all contribute to the insect host range of the bacterium.

SAQ 8.16

List at least two possible explanations for the absence of toxicity of *B. thuringiensis* against mammalian species.

processing of toxin

In relation to molecular plant breeding, one more property of the crystal proteins should be discussed. During conversion from protoxin to toxin the C terminal half of the protein and a short sequence from the N terminal end are removed to produce the toxin. When a crystal protein gene is truncated at its 3′ end such that the entire coding sequence for the toxic fragment remains intact, the resulting protein product is still fully toxic when ingested by a sensitive larva. Also, in *Escherichia coli*, the bacterium in which genetic engineering has been carried out, the truncated product is synthesised in quantities similar to those of the original protoxin. Apparently, the 3′ end of the gene can be omitted. The next part of this section will show the importance of this finding.

One of the disadvantages of the use of *B. thuringiensis* as a bio-insecticide is the relatively short survival time of crystals of *B. thuringiensis* toxin in the field. This necessitates repetitive costly applications of the bacterium.

SAQ 8.17

Do you think *B. thuringiensis* efficiently grows in the field? Why?

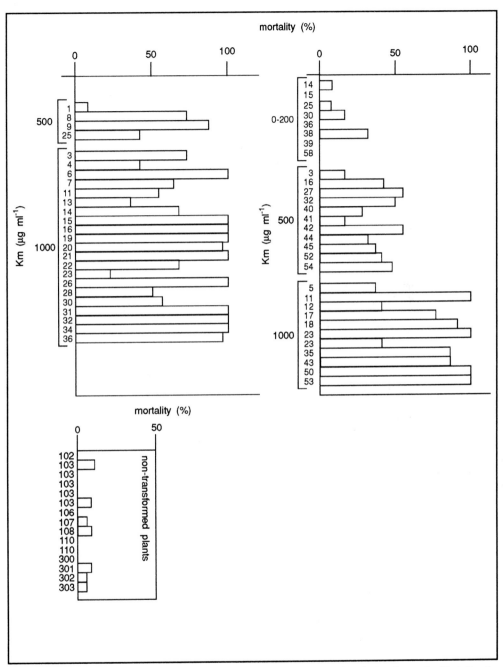

Figure 8.10 Insect toxicity of transgenic tobacco plants expressing the *B. thuringiensis* protein. Mortality of *Manduca sexta* larvae after 6 days of feeding on tobacco leaves is shown. The serial number of each plant is given below each column, plants are grouped according to the concentration (in µg/ml) of kanamycin (Km) to which they were resistant. This directly reflects the amount of *B. thuringiensis* protein produced by the plant. (After Vaeck *et al* (1987) Nature 328, 33-37).

Despite the variety of *B. thuringiensis* proteins, it was recognised that, at least in some isolates, the toxic activity was located in a single protein, in other words that the toxicity relied on a monogenic trait. Thus, it is not surprising that several research groups have attempted to generate insect resistant transgenic plants based on the transfer of *B.thuringiensis* crystal protein genes. The first group to succeed in such an attempt was the Ghent-based group of Plant Genetic Systems, a biotechnology company. They generated transgenic tobacco plants that had incorporated into their genome either an intact crystal protein gene or a derivative of this gene from which the 3' end had been deleted. Moreover, a derivative was used in which a truncated gene as above had been fused to the gene coding for kanamycin resistance (NPTII), such that a fusion protein resulted containing crystal protein and neomycin phosphotransferase sequences.

intact crystal protein gene restricts its own expression

Surprisingly, none of the plants containing a copy of the intact crystal protein gene appeared to be toxic to the test insect, the lepidopteran tobacco horn worm *Manduca sexta*. On the other hand, several plants containing one or more copies of the truncated gene version were fully resistant to larvae of this insect. It was shown that the crystal protein levels in plants containing intact gene copies were consistently lower than in plants containing truncated gene copies. Apparently, the coding sequence of the *B. thuringiensis* crystal protein gene somehow interferes with its expression in plant cells. Even the expression levels obtained in the plants containing truncated gene copies were lower (0.004 - 0.02%) than might be expected for the plant promoters used (the 35S CaMV and T-DNA 2' promoters). At present no clear explanation for this phenomenon is available. This result has been observed with at least one more gene (excessively low expression of the gene coding for human serum albumin in potato tuber).

NPT genes to select for crystal protein production

It was shown that the fusion protein consisting of crystal protein and NPTII domains still contained both activities. Moreover, in transgenic plants a strong correlation between the antibiotic resistance level and the degree of insect resistance was observed (see Figure 8.10). This enabled the selection in an early stage of the regeneration procedure of those transgenic shoots with highest kanamycin resistance levels in the search for highly insect resistant plants. More generally, the use of a so called NPTII fusion may thus facilitate an easy selection of plants showing highest expression of proteins responsible for a desired trait.

SAQ 8.18

Can NPTII fusions be used for every gene to be transferred? What limitations can be expected? (Think about the interaction of the two protein domains).

Insect resistant plants belonging to tobacco, tomato, potato, sugar beet, rapeseed and poplar species have been obtained by several different research groups. However, resistance is limited by the host range of the crystal protein being produced in the transgenic plants. Several different crystal protein genes will have to be exploited for an efficient control of many important plague insects.

SAQ 8.19

What makes the range of resistance of the transgenic plants narrower than that of many bacterial isolates?

multiple crystal protein gene transfer

The potential emergence of insect populations which have become insensitive to a given *B. thuringiensis* crystal protein expressed in transgenic plants will have to be carefully monitored. In the meantime strategies to avoid such an emergence, like the simultaneous transfer of two genes of which the products both control the same insect species should be worked out. After all, the monogenic nature of the resistance trait, coupled to a high and continuous selection pressure in transgenic crop fields might form an optimal condition for the emergence of crystal protein resistance.

Insect resistance based on proteinase inhibitors

proteinase
inhibitor

Following the first reports on insect resistant plants, in which resistance had been achieved by the use of crystal protein genes, the successful engineering of tobacco plants containing the coding sequence for a trypsin inhibitor from cowpea (*Vigna unguiculata*) was reported by a research group from the University of Durham. For several decades the abundant presence in plant storage organs (seeds, fruits, tubers) of proteins which inhibit the activity of proteolytic enzymes, proteinases such as trypsin and chymotrypsin had been demonstrated. Potato tubers for instance contain at least six different types of these proteins, called proteinase inhibitors, amounting up to 20% of total cell protein. Young tomato fruits also show high levels of proteinase inhibitors, partly showing homology to the inhibitors isolated from potato tubers. Most of them are encoded by small gene families, consisting of highly homologous but nonidentical members. In more recent years it has been shown that the appearance of similar or identical proteins can also be induced by wounding or insect attack in other parts of the plants. Proteinase inhibitors act by tightly binding to the reactive site(s) of a proteinase.

Insects often contain large amounts of trypsin and/or chymotrypsin in their intestinal tract for food protein digestion, whereas few plants contain these enzymes. Therefore, it is generally believed that proteinase inhibitors play some role in the plant defence system. By producing these inhibitors either constitutively as prevention in storage organs or after wounding, as in the leaves of several plant species, plants may protect themselves against large scale insect damage through interference with the insect metabolism.

SAQ 8.20

Immature tomato fruits contain high levels of proteinase inhibitors but these levels decrease greatly at fruit ripening. What might be the biological function of this decrease?

efficiency of
cowpea trypsin
inhibitor as
protectant
against insects

This feature has been exploited in the example mentioned above in which transgenic tobacco plants producing cowpea trypsin inhibitor up to 0.8% of the total amount of cell protein were engineered. These showed partial or complete resistance to a considerable array of insects. A comparable result was obtained when a cDNA copy for a potato inhibitor II transcript was used. Although in both cases the reported array of sensitive insect larvae is larger than in the case of the plants engineered with *B. thuringiensis* crystal protein genes, the level of protection is less.

Several more proteinase inhibitors are good candidates to provide similar insect resistance. Also, plants containing both a crystal protein gene and the coding sequence for a proteinase inhibitor might exhibit synergistic effects and thus higher resistance than the plants expressing only one of these proteins, and chances of the development of insensitive insect populations might thus be diminished.

Π Can you think of any risks which might be associated with introduction of proteinase inhibitors? (Think about the use the plant is put to).

If constitutively expressed it could enter fruit or vegetables eaten raw (cf proteinase inhibitors under developmental control in tomato, sited in SAQ 8.20). Alternatively they might also be present in pollen and nectar collected by pollinating insects.

8.4.3 Herbicide resistance

non specificity
of herbicides

The use of herbicides amounts to one third of total pesticide use and more than a hundred different herbicides can be commercially purchased. Of course, ideal herbicides should combine high effectiveness with low toxicity and efficient biological degradation. However, most herbicides interact with general plant specific processes like photosynthesis and amino acid synthesis pathways which are alike in crops and weeds. This renders these herbicides non-selective. Many new and powerful herbicides that are easily degraded and show extremely low toxicity to mammals share these properties. If crops can be engineered such that they become herbicide resistant, these herbicides could still be exploited and expensive screening programs to obtain more selective herbicides exhibiting comparable properties could be avoided. Table 8.1 lists some examples and indicates their mechanism of action.

Example of herbicides	Active ingredient	Site of action
Round up R, Herbiace R, Mustro R, Sting R	glyphosate	5-enolpyruvyl-3-phosphoshikimate synthase
Oust R, Sting R	metsulphuron-methyl	acetolactate synthase
Glean	chlorsulfuron	acetolactate synthase
Buctril R, Buctrilin R	bromoxynil	quinone-binding
Basta R	phospinothricine	glutamine synthase

Table 8.1 Action of different herbicides on steps in photosynthesis and biosynthesis of amino acids.

economic and
environmental
consequence
of producing
herbicide
resistant plants

Here the basic strategies used for the construction of transgenic plants will be discussed. The reader should be aware of the need for a thorough risk assessment, in which the risks for transfer of genes responsible for herbicide resistance from the engineered crops to sensitive weeds are estimated on the basis of experimental studies. We dealt with these issues more thoroughly in Chapter 2 so therefore we will not enlarge on them here. Also, the reader should bear in mind that no judgement is given on the social aspects of the introduction of herbicide resistant plants. These aspects include the growing economic dependence on few manufacturers, selling a package of herbicides and herbicide resistant seeds, and a potentially excessive use of the herbicides involved facilitated by the high field resistance levels of the transgenic plants.

Resistance by a mutated target gene

herbicide
interaction with
a metabolic
pathway

Herbicides interfering with biochemical pathways do so by interacting with one of the enzymes catalysing this pathway (the target enzyme). Spontaneous or selected mutant target enzymes have been described that have become largely insensitive to the herbicide while still able to fulfil their normal activity. If the gene coding for a mutated enzyme is transferred into a sensitive plant, the mutated enzyme can take over the role of the resident wild-type enzyme.

SAQ 8.21

Will the resident, original gene still be expressed in the transgenic plant?

Transgenic plants have been produced that show resistance to herbicides interfering with plant specific pathways for amino acid synthesis. Examples of such herbicides are sulphonyl urea compounds (eg Glean) which interact with acetolactate synthase, a key enzyme in the biosynthesis of the branched chain amino acids valine, leucine and isoleucine, and glyphosate (eg Roundup) interacting with the 5-enolpyruvylshikimate-3-phosphoshate synthase involved in the biosynthesis of aromatic amino acids. Mutant genes have been isolated from different sources. For instance, a mutant gene from *Salmonella typhimurium* showing resistance to glyphosate has been isolated by selection for resistance against the herbicide in the bacteria. The mutant enzyme shows an amino acid substitution (proline to serine) leading to decreased affinity of glyphosate for the enzyme.

SAQ 8.22   Why are such mutants selected in bacteria?

Since the acetolactate synthase genes appeared to be highly conserved, it was possible for the isolation of the plant gene members to be accomplished by hybridisation of a heterologous acetolactate synthase gene from yeast to plant genome libraries of wild type and sulfonylurea resistant cell lines of tobacco and *Arabidopsis thaliana*. A highly resistant tobacco line showed two amino acid substitutions in its acetolactate synthase.

Tolerance by overproduction of the target enzyme

By introduction of extra copies of the gene coding for the target enzyme, preferably under the control of a strong promoter such as the 35S CaMV promoter, sensitivity to herbicides has also been overcome. Herbicide tolerance is caused by overproduction of the target enzyme. Following this strategy glyphosate tolerant petunia and tobacco plants showing over production of 5-enolpyruvylshikimate-3-phosphate synthase have been engineered. A similar result has been obtained by selection for glyphosate tolerant petunia and alfalfa cells in tissue culture, in which a natural amplification of the genes coding for the target enzymes had taken place.

SAQ 8.23   Explain the difference between resistance as described in the previous section and tolerance as described here.

Resistance by introduction of a gene coding for a detoxifying enzyme

bialophos

phosphino-
thricine

An antibiotic with herbicidal activity is produced by the bacterium *Streptomyces hygroscopicus*, it is called bialophos. It consists of two alanine residues and a glutamic acid analogue called phosphinothricine. Phosphinothricine, which is generated from bialophos by the action of peptidases, strongly inhibits glutamine synthetase which catalyses the conversion of glutamic acid to glutamine. Inhibition of glutamine synthetase blocks the metabolism of ammonia which depends on this pathway, leading to highly toxic levels of this essential compound in the plant. The bialophos producing bacterium *S. hygroscopicus* itself produces, for obvious reasons, a detoxifying enzyme acetylating the free amino group of phosphinothricine. This enzyme is called phosphinothricine acetyltransferase or simply PAT. The structures of these compounds are given in Figure 8.11.

SAQ 8.24   Why is the production of a detoxifying enzyme by *S. hygroscopicus* obvious?

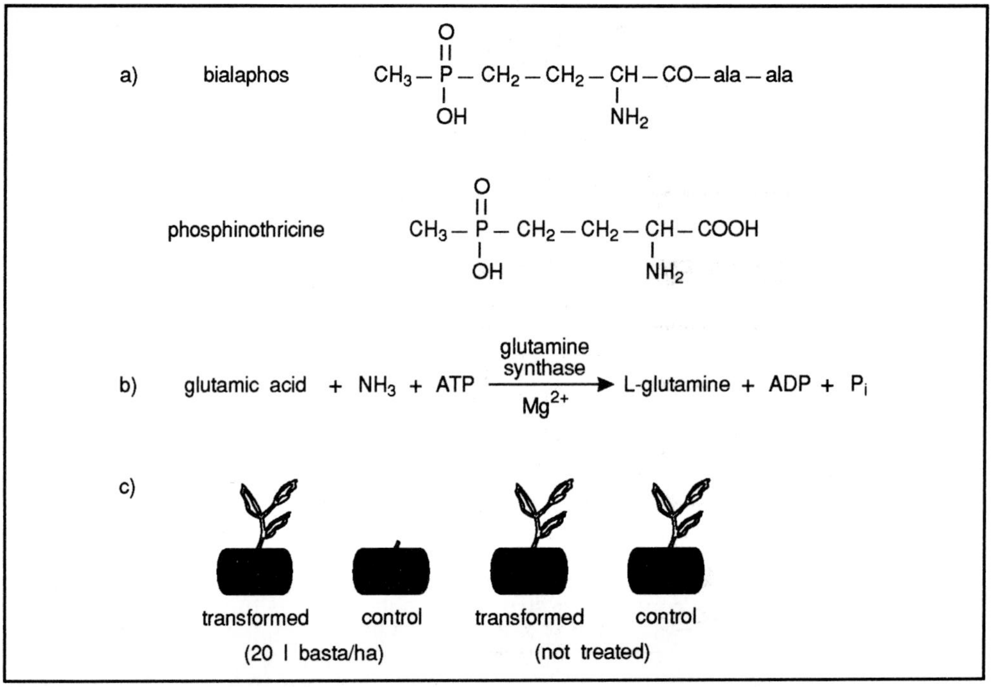

Figure 8.11 Chemical structure of bialaphos and phosphinotricine (panel a)). These compounds interfere with the metabolism of L-glutamic acid by inhibition of the enzyme glutamine synthase (panel b)). Panel c) shows transgenic, herbicide resistant and control tobacco plants treated or not treated with the herbicide basta (see text for details).

small amounts of PAT provide resistance

The acetylated reaction product appears to be non-toxic to the organism and for plants. The gene coding for PAT has been isolated, fused to the 35S CaMV promoter and transferred to tobacco cells which were then screened for resistance against phosphinothricine. After regeneration, fully resistant plants emerged showing the presence of the acetylating enzyme PAT. Levels of 0.001% of total cell protein appeared to be sufficient for a fully resistant phenotype under normal field applications of phosphinothricine.

SAQ 8.25

Is the usual selection for kanamycin resistance a necessary part of this strategy? Indicate why.

Following an identical strategy resistant potato, tomato, rapeseed, sugar beet and poplar plants have been obtained. These resistant plants were the first to be investigated in European field trials in 1987 and 1988. The dominant resistance trait appeared to be stably inherited under field conditions.

SAQ 8.26

Why is resistance dominant?

environmental fate of acetylated phosphinothricine

The fate of the acetylated phosphinothricine in the plant and, later on, in the environment is still unclear. If the product of the detoxification reaction is only marginally metabolised, this will lead to its accumulation with yet unknown consequences, and this in turn might pose a serious threat to the application of phosphinothricine in transgenic crops. We can summarise the strategies for producing herbicide resistance by the little flow diagram.

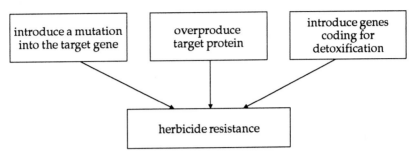

8.4.4 Bacterial resistance

small peptides involved in defence

As yet no transgenic plants showing resistance to bacterial pests have been obtained. However, the first successful steps on the road to bacterial resistance in plants have been taken. Small peptides of different origin have been implicated in potent host defence mechanisms against bacterial invasions. The amino acid sequences of some of these peptides, and their bactericidal spectra have been elucidated. Furthermore, the genes coding for these peptides, or cDNA copies derived from their mRNA transcripts, have been isolated. Alternatively, synthetic genes based on the amino acid sequences have been constructed. In due course, transgenic plants producing these peptides will be generated. These plants can then be screened for their resistance against bacterial pests.

Bactericidal peptides whose expression in transgenic plants is presently being aimed at, fall into two different groups. One group consists of proteins involved in humoral (cell free) immunity in insects. Humoral immunity can be induced in insects by an injection of either live, non pathogenic bacteria or heat killed pathogens. It is based on the expression of a set of proteins, normally not produced in these animals. Among 15 different immune proteins produced by the giant silk moth *Hyalophora cecropia* are the

cecropins

cecropins. They are small basic proteins of approximately 40 amino acid residues in size which exhibit a broad toxicity spectrum against both Gram positive and Gram negative bacteria at promisingly low doses. Cecropins are the products of three related genes that have originated by gene duplication. The presence of similar humoral, anti-bacterial peptides, even smaller in size (28 amino acid residues) known as apidaecins have been

apidaecins

demonstrated in the honey bee *Apis mellifera* after challenge with sublethal doses of live bacteria. Plants producing cecropins or apidaecins are being produced.

thionins

The other group of anti-bacterial proteins exploited in molecular plant breeding are the thionins. These small basic proteins (45 to 50 amino acid residues in size) were first isolated from the endosperm (seed tissue) of different cereals (*Gramineae*) as the causal agent of the growth inhibition of fermenting yeast. Elucidation of cDNA sequences demonstrated that mature thionins are the processed products of larger precursor molecules. These consist of a signal peptide of 25 to 30 residues in size necessary for correct membrane transport of the precursor thionin molecule, followed by the mature basic peptide and an acidic terminal part (approximately 65 residues in size) of an unknown function. It is suggested this part plays a role in protection of the host cell against thionin action. We can thus draw the structure as:

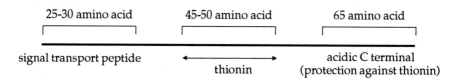

Recently it was found that other members of this peptide family occur in leaves, both in the cell wall and in the vacuole. Thionins seem to exert their toxic effect at the level of the cell membrane of the pathogen. Thionin coding sequences can be used in resistance breeding of solanaceous crops as potato and tomato, species that lack these proteins.

| **SAQ 8.27** | What is the likely explanation for the abundant presence of thionins in the seed endosperm? |

All peptide groups mentioned show toxicity against *Clavibacter michiganense pathovars* (pathological varieties), the causative agents of ringrot disease in potato and wilting disease in tomato and against *Xanthomonas campestris* pathovars, causing leaf spot disease in tomato. Phytopathogenic *Erwinia* and *Pseudomonas* species were also shown to be sensitive to some of these peptides.

lysozyme

attacins

Apart from cecropins, challenged pupae of *H. cecropia* also produce a lysozyme (an enzyme which degrades bacterial cell walls) and attacins. Attacins have a molecular weight of 20 000 daltons and are comprised of six different forms, basic and acidic variants encoded by two related genes. It was shown that cecropins, lysozyme and attacins act synergistically (ie they enhance each others activity). Using these products together for resistance breeding might even make it more difficult for the bacterial pathogen to compete.

8.4.5 Pathogenesis related proteins

The process of disease induction by infectious micro-organisms and viruses (pathogens) is called pathogenesis. Pathogenesis related proteins (PR proteins) are those proteins that can be induced by a pathogen infection.

hyposensitive response

necrosis

Reactions of the host plant to a given pathogen may vary widely. Some plants show no visible reactions, a successful infection does not take place and plants remain healthy. Other plants show severe symptoms upon infection and eventually die. The so called hypersensitive response (HR), characterised by a quick, localised necrosis (cell death) at the infection site, falls somewhere inbetween. A hypersensitive response is often accompanied by only a limited spread of the pathogen to non-infected parts of the plant. Therefore HR can be considered as a host defence reaction.

| **SAQ 8.28** | What might be the function of a localised necrosis for the plant? |

pathogenesis related protein action

A hypersensitive response involves expression of a series of specific plant genes which on the one hand lead to death of a limited number of cells at the infection site (localised necrosis) and on the other hand to the appearance of many new proteins, both close to the infection site and in distant parts of the plants. Many different pathogens and even some chemical compounds can all induce similar proteins which may render the host resistant to a second attack by the same agent. They may also protect the plant from entirely unrelated pathogens. Bacterial infection inducing HR in tobacco may result in resistance against tobacco mosaic virus and vice versa.

lignins and phytoalexins

Intracellular proteins induced in a hypersensitive response near the site of infection include key enzymes in the biochemical pathways for the production of aromatic compounds such as lignins and phytoalexins. Lignins are incorporated into cell walls whereas phytoalexins show antimicrobial activity. Both contribute to inhibition of the spread and replication of microbial pathogens. Simultaneously, newly induced proteins are secreted into the extracellular space at many different sites in the infected plant. These proteins are the so called pathogenesis related (PR) proteins.

SAQ 8.29

Why is secretion of these proteins into the extracellular space useful?

---

role of PR proteins in host resistance

PR proteins have been demonstrated in over twenty different species. Of these, tobacco has been most thoroughly studied. This species shows the presence of at least ten PR proteins, many of which have been purified. Some PR proteins show enzymatic activity, such as chitinase or β-1,3 glucanase activity. Other proteins showed homology to an amylase/proteinase inhibitor from maize. The substrates for these enzymes, chitin and β-1,3 glucan, are important compounds of the cell walls of many fungi and it is known that proteinase inhibitors may cause retardation of insect larval growth. It is therefore generally believed that PR proteins play a role in plant resistance. Recent data show the existence of basic homologues probably located in the cellular vacuole, to the acidic PR proteins isolated from the extracellular space. Like their acidic counterparts, the basic homologues may be induced in stems and leaves, but they are constitutively present in plant roots.

Another member of the PR protein family is a glycine rich protein consisting of 25% glycine residues and probably anchored in the plant cell wall for protective purposes. Isolated and cloned PR protein genes might provide new tools for engineering resistance in transgenic plants. At present, no such plants are yet developed.

8.4.6 Summary of developing resistance and tolerance by genetic engineering

In the sections above, we have illustrated the use of genetic engineering to provide plants with protection against disease, parasites and herbicide toxicity. The genes used have come from a wide variety of sources including viruses, bacteria, insects and plants. Carry out the in-text activity below, it will help you to remember the sources of genes and strategies involved.

Π Below we have drawn an outline scheme of the strategies used to provide protection against infections, predators and parasites and to develop herbicide resistance. Draw out this scheme and write on it, the sources of the genes that are used to provide protection. (You will need to look back over Section 8.4).

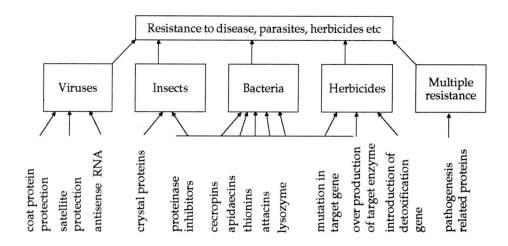

8.5 Other transferred traits

In this section we discuss the successful manipulation of entirely different traits and thus show the potential application of molecular biological techniques in plant breeding for many different and diverse traits.

8.5.1 Flower colour in petunia

Manipulation of flower colour has long been one of the central issues in the breeding of ornamental plants. However, the usual limitations of classical plant breeding such as species barriers and the random outcome of crosses explain the limited progress in this field. It is for this reason that blue tulips or blue roses can not be obtained. Recent advances in the isolation of genes and the transformation of ornamental crops now make it possible to alter an aesthetic property such as flower colour in a directed way.

flavonoid pigments

The most common flower pigments are flavonoids, aromatic compounds possessing three different ring structures. Flavonoids can generate a colour spectrum varying from red to blue. The biosynthetic pathway has been studied thoroughly, especially in maize and petunia. The aromatic amino acid phenylalanine forms the precursor for flavonoids and the second conversion step common to all flavonoids is catalysed by chalcone synthase.

use of antisense genes

gene conservation

Researchers at the Free University Genetics group in Amsterdam interrupted this pathway in petunia by expressing the inverse orientation (antisense) of the chalcone synthase gene. Upon transcription of the antisense DNA, an antisense RNA is formed which probably base pairs with the complementary mRNA coding for the chalcone synthase. As a result, the mRNA can not be translated and the phenotypic effect will be that of a mutation. Since the substrates for chalcone synthase are colourless, a fainter colour or even pure white would be expected, which is indeed observed. It was shown that the antisense petunia chalcone synthase gene is also effective in heterologous plants such as tobacco, probably because the gene involved is sufficiently conserved between at least these distantly related species.

The power of the antisense technique (that shows homology to naturally occurring antisense regulation in bacterial systems) resides in the possibility of blocking individual steps in a complex biosynthetic pathway, thus accumulating an intermediate metabolite.

diversification
of metabolic
products

The actual first case of novel pigmentation of petunia flowers was reported in 1987. A maize gene coding for the enzyme dihydroflavonol 4-reductase was transferred to petunia. The maize enzyme catalyses the production of three different flavonoids whereas the resident petunia homologue can only use two different substrates. Expression of the maize gene led to the appearance of brick red petunia flowers resulting from the presence of the third reaction product.

SAQ 8.30

Describe the fundamental difference between the two strategies described above which both led to the production of new flower colours.

8.5.2 Starch composition of the potato tuber

granule bound
starch synthase

mosaic of cell
types

Starch in potato tubers consists of 20 to 25% amylose (linear polymer of glucose linked by α-1,4 links) and 75 to 80% amylopectin, a branched form of amylose showing some α-1,6 links between glucose residues. An important enzyme involved in the synthesis of amylose in starch is the granule bound starch synthase (GBSS). The starch granules are present in the amyloplasts (related to chloroplasts) in the tuber cells. This catalyses the chain elongation of amylose by addition of glucose residues. In an attempt to inhibit GBSS activity in potato tubers and thus affect the amylose/amylopectin ratio in the cells, the antisense strategy was applied. Recombinant T-DNA constructs containing part of the GBSS cDNA in the inverse orientation were introduced in potato by the use of *Agrobacterium rhizogenes,* which causes the formation of hairy roots. These roots were stained with a dye that stains pure amylopectin reddish brown and amylose containing starch blue. A pattern (mosaic) of blue and reddish brown cells was observed.

Regenerated plants from these hairy roots were induced to form tubers and starch of these tubers was analysed. Some tubers were observed that showed a considerably lower amylose content compared to wild type tubers judged from the colour. These tubers exhibited strongly reduced GBSS activity levels (up to 85% reduction). By these experiments it is shown that it is possible to create new potato starches with altered contents, which is potentially useful for the potato processing industry.

SAQ 8.31

Each tuber exhibited a homogeneous staining pattern. Do you think the origin of the tuber is unicellular or multicellular?

8.5.3 Tomato fruit softening

effects of
antisense gene
for poly-
galacturonase
on tomato
ripening

Polygalacturonase is the major cell wall degrading enzyme of tomato fruit and is believed to play a key role in the fruit softening process. It is only synthesised in ripening fruit and not in other developmental stages of the fruit nor in other tomato plant organs. The enzyme has been isolated from mature tomatoes and shown to be associated with the fruit cell walls. Again, a construct containing part of the cDNA sequence in inverse orientation under the control of the 35S CaMV promoter was produced and transferred to tomato. Transgenic plants were induced to flower and left to set fruits. The fruits of the transgenic plant showed dramatically postponed softening as compared to normal control fruits. Softening only taking place up to four weeks later. Levels of polygalacturonase were severely reduced.

This example clearly offers possibilities for application, since fruits produced from these transgenic plants will have prolonged shelf life and hence improved quality.

Summary and objectives

In this chapter we have dwelled on developing resistance to infection and parasites. We have provided a broad sweep of this respect of plant biotechnology. In the next chapter we focus onto a particular case of plant protection which provides an alternative strategy for protecting plants. This will provide you with an in-depth appreciation of what is involved in applying biotechnology to plant protection. This chapter has covered two main issues:

- the techniques available for introducing new genes into plants using molecular biological procedures;

- the choice of genes to be used for such genetic manipulation and their consequences on plant phenotypes and performance.

Now that you have completed this chapter you should be able to:

- draw a flow diagram which describes the sequence of events which leads to the transfer of T-DNA from *Agrobacterium* into plant cells;

- assign roles to vir and T-genes in T-DNA transfer and tumour development;

- describe strategies for selecting plant cells transformed by T-DNA and be able to explain the differences between a selectable gene and a reporter gene;

- list a variety of plant promoters and describe the factors which influence their activities;

- explain how PEG, micro- and macro-injection of DNA may be used to transform plants which are not transformed by agrobacterial T-DNA;

- list sources of genes which may be used to protect plants from biological and chemical damage;

- describe strategies for using non indigenous plant genes for producing new phenotypic characteristics particularly with reference to disease infection;

- explain how antisense mRNA can be made and how it has been used to modify phenotypic characters.

Case study: Baculoviruses - genetically engineered insecticides

9.1 Introduction and scope   222

9.2 Baculovirus biology   223

9.3 Baculovirus genetics   228

9.4 Genetic engineering of baculoviruses expression vectors   231

9.5 Strategies for the engineering of baculovirus insecticides   233

9.6 Production of (recombinant) baculoviruses   238

9.7 Application   239

9.8 Conclusions   240

Summary and objectives   241

Case study: Baculoviruses - genetically engineered insecticides

9.1 Introduction and scope

About one third of the loss of agricultural products in the world is caused by pests and diseases. Pest control is therefore of eminent importance to meet the needs for sufficient food for the world's population and raw material for industrial applications. The use of chemical pesticides has been a world wide response to combat pests and diseases. However, the environmental hazards associated with the use of such chemicals and increased public concern have led to an intensified search for alternative strategies, including the use of biological agents.

Insect pests can be controlled through release of naturally occurring biological entities such as parasites, predators and pathogens in the infested ecosystem. Microbial agents such as bacteria, fungi and viruses are capable of controlling insect pests, genetic engineering of these pathogens may make them more effective. As an example of the application of this new technology the perspectives for genetic engineering of viral insecticides is discussed in this chapter. These new viral pathogens should combine the property of a restricted host range with improved insecticidal activity. Of particular value are baculoviruses that cause fatal diseases in insects. This chapter focuses on these viruses.

The major aims in the engineering of baculovirus insecticides are:

* increase of speed of action, in particular the cessation of feeding;

* enhancement of virulence, in particular against old larvae;

* extension of a defined host range in order to make their production commercially attractive;

* improved persistence in particular with regard to UV sensitivity in the field.

With these aims in mind, we will provide some background information concerning baculoviruses before examining how we can genetically engineer the virus to have more desirable properties.

SAQ 9.1	1) What are the major advantages of biological control agents over the use of chemicals?
	2) Which of the aims of genetic engineering of baculoviruses will be the most difficult ones to achieve? What could be an alternative solution to this problem?

9.2 Baculovirus biology

9.2.1 Baculoviruses

epizootic

Baculoviruses are viral pathogens, that cause fatal diseases in insects, mainly in members of the orders *Lepidoptera* and *Diptera*. They are highly specific and can attack large numbers of hosts over a short period (ie epizootic) thereby reducing the size of insect populations in nature. Therefore, baculoviruses are recognised as potential biological control agents of insect pests in agriculture and forestry as alternative to chemical insecticides. Over 500 baculoviruses are known (Table 9.1).

Insect orders/ class	Cytoplasmic polyhedrosis viruses	Entomopox viruses	Baculoviruses
Insecta			
Lepidoptera	191	24	456
Diptera	37	9	27
Hymenoptera	6	3	30
Coleoptera	2	14	9
Neuroptera	2	-	-
Trichoptera	-	-	2
Orthoptera	-	9	-
Decapoda	-	-	2
Crustacea	-	-	2
	238	59	528

Table 9.1 Occluded viruses of arthropods.

∏ Using Table 9.1, which group of arthropods are subject to the greatest diversity of baculoviruses?

specificity of baculoviruses

Baculoviruses are only found in the arthropod kingdom, mainly from insects and a few from crustaceans. They are usually host specific and do not affect non-target species or other organisms, including man. They are a natural element in the environment causing viral epizootics and have a long safety record as control agent of insect pests. They are compatible with the use of other pesticides (eg fungicides) and can be applied onto crops with conventional spray equipment. As with all viruses it takes some time for the disease to develop. In the case of baculoviruses, four to seven days go by before the first symptoms are apparent (Figure 9.1).

∏ Does the delay in baculovirus action have any important practical consequences for growers? Examine Figure 9.1 and think what is happening during the first three days after infection - have the larvae stopped feeding?

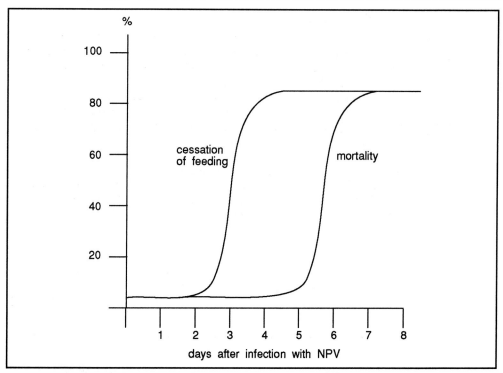

Figure 9.1 Cessation of feeding and mortality of insects versus time after baculovirus infection. (NPV stands for nuclear polyhedrosis virus - see text).

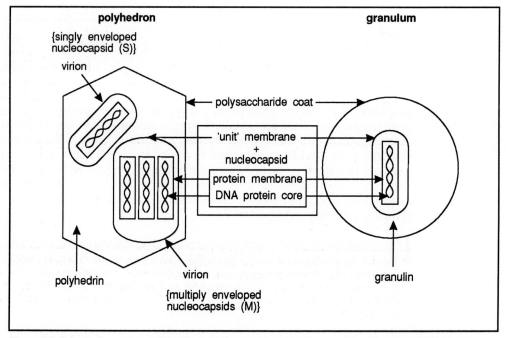

Figure 9.2 Schematic representation of baculovirus occlusion bodies: a polyhedron (from NPVs) and a granulum (from GVs).

9.2.2 Structure

granulosis and polyhedrosis viruses

Baculoviruses are characterised by the presence of rod shaped (baculum = rod) nucleocapsids, that are enveloped singly or in bundles by a membrane (Figure 9.2). The virus particles are usually embedded singly (granulosis viruses = GV) or in large numbers (nuclear polyhedrosis viruses = NPV) into large protein capsules, called occlusion bodies (OBs), granula or polyhedra. The OBs are surrounded by an envelope, consisting of protein and carbohydrate. OBs vary in size between 1-15 μm (NPVs) and 0.1-0.5 μm (GV) in diameter and can be observed by light microscopy. The OB protects the virus particles against decay in the environment.

occlusion bodies

granula polyhedra

Occlusion of virus particles is not an exclusive property of baculoviruses. Cytoplasmic polyhedrosis viruses, belonging to the reovirus family and entomopox viruses, classified in the poxvirus family, are also occluded. Representatives of these two families found throughout the animal kingdom. Only baculoviruses are restricted to arthropods.

polyhedrin or granulin

The major constituent of these OBs is a single protein, polyhedrin or granulin, with a subunit molecular weight of approximately 30 000 dalton (30 kDa). The protein sequence of granulins and polyhedrins is highly conserved among all baculoviruses. There is no homology with the occlusion body proteins of cytoplasmic polyhedrosis and entomopox viruses.

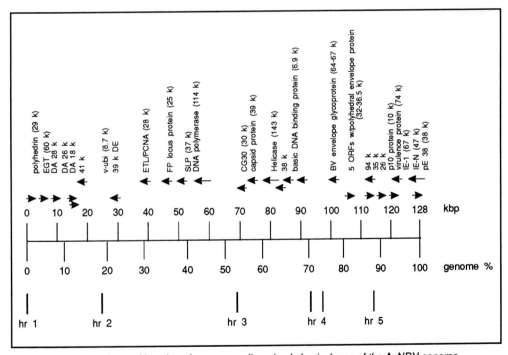

Figure 9.3 Organisation and location of genes on a linearized physical map of the AcNPV genome. Numbers (in kDa) refer to appropriate molecular weights of the proteins coded for by these genes. Arrows indicate the approximate size and 5'-3' orientation of the coding regions. The homologous repeat regions (hr1-hr5) are indicated as vertical bars (modified from Blissard and Rohrmann, 1990, Annual Reviews of Entomology 35,127).

baculovirus genome

The genetic information of baculoviruses is contained in a double stranded, circular DNA molecule. The viral DNA varies in size between 80 and 200 kilobase pairs (kbp) and is able to code for more than sixty average sized proteins. Physical maps of various baculovirus DNAs have been established, the most detailed one being of the prototype baculovirus *Autographa californica* nuclear polyhedrosis virus (AcNPV - 130 kbp). (Figure 9.3). About twenty five genes have been mapped on the genome, including the gene for polyhedrin (Figure 9.3). The nucleotide sequence of about ten genes is known.

∏ Use the description above to work out roughly how many nucleotides can be placed in order in the AcNPV genome.

Your calculation can only be approximate. This viral genome consists of 130 kbp. 10 genes (ie those whose nucleotide sequence is known) represents about $\frac{1}{6}$ (10/60) of the genome, ie about 21 kbp. Thus we can write a nucleotide sequence for about 21 kbp.

9.2.3 Replication of the virus

biphasic cycle

Follow Figure 9.4 while you read the following description. In nature, OBs enter the larvae via ingestion of contaminated food and dissolve in the midgut as a consequence of the local alkaline conditions (Figure 9.4 A). The released virus particles infect midgut cells and replicate in their nuclei (Figure 9.4 B, C and D). Progeny virus particles (non-occluded virus particles = NOV) bud through the midgut cell membrane and are released in the insect hemolymph (Figure 9.4 H, I and J). These NOVs are infectious for insect cells and are transported via the haemolymph to other organs of the insect, such as the fat body (Figure 9.4 K). In cells of these organs, a second round of NOV replication occurs followed by production of virions which are occluded in the nucleus in newly synthesised OBs (Figure 9.4 E, F, G). At the end of the infection, the cells lyse and release the OBs. This biphasic replication process can be mimicked in cultured insect cells and this allows a detailed cytopathological, molecular biological and genetic study of baculovirus replication. The replication of *Autographa californica* NPV (AcNPV) in *Spodoptera frugiperda (Sf)* cells is the best studied example.

9.2.4 Gross pathology

symptoms and time course of disease development

The initial signs of baculovirus infection appear about three days after infection. In the early stages of the disease, haemocytes show enlargement of the nucleus with OBs occupying the entire nucleus and giving a whitish discoloration of the haemolymph. Usually, baculoviruses can infect a broad range of tissues, including fat body, tracheal matrix and muscles. Effects on in the rate of development are also apparent at this time. Healthy larvae continue to develop, infected larvae begin to lag behind. Infected larvae exhibit the classical signs of baculovirus infection including loss of appetite and lethargy. In more advanced stages of disease, six to seven days after infection, the larval body usually develops a creamy yellow colour due to accumulation of OBs in the infected tissues. Subsequently, the larvae become flaccid and hang, by the prolegs, from leaves or branches in a characteristic inverted position, or detach from the plants and fall to the soil. Larvae die, dependent on the size of the dose, three to eight days after the initial signs of infection after which the cuticle ruptures liberating masses of OBs (about 10^9 per larva). The slow speed of action accompanied by the late cessation of feeding (Figure 9.1) is a major limitation in the use of these pathogens to control insect pests, particularly in crops with low damage thresholds.

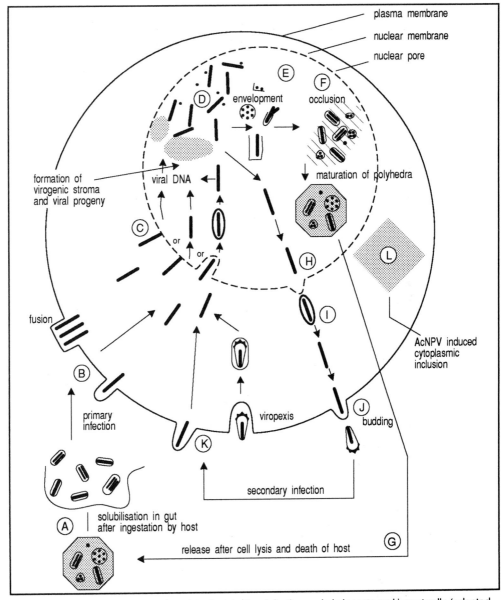

Figure 9.4 Schematic representation of the baculovirus infection cycle in insects and insect cells (adapted from Blissard and Rohrmann, 1990, Annual Reviews of Entomology 35, 127). See text for details.

SAQ 9.2

1) Make a list of reasons why baculoviruses may be host specific.

2) Explain what could be the evolutionary significance of the conservation of polyhedrin and granulin genes in baculoviruses.

9.2.5 Resistance

mature larvae
are more
resistant than
young larvae

Information on active defence mechanisms in insects against virus infection is limited. The exoskeleton (cuticle) provides most of the protection. Entry of baculoviruses via the digestive tract bypasses this defence. The defence system of insects does not include an active immune response in the form of antibodies like that found in vertebrates. Antiviral proteins are observed, but their role is unclear. The major form of resistance is of a genetic nature, as instar larvae become progressively less susceptible to viral infection with age (Table 9.2). Hence, in order to reach 50% mortality (LD_{50}) about 10^4 times more OBs are required to kill fifth instar larvae as compared to first instars (NB instar = larval stage between moults). Increase of virulence is therefore an important aim in the genetic improvement of baculoviruses.

		95% fiducial limits	
Instar	LD_{50}	lower	upper
L1	4	3	6
L2	3	2	4
L3	39	21	94
L4	55	35	180
L5	11637	6210	32544

Table 9.2 Mortality (LD_{50}) of *Spodoptera exigua* NPV, expressed as numbers of occlusion bodies required for LD_{50}, against the five larval instars of beet armyworm, *Spodoptera exigua*.

∏ Can you produce a hypothesis as to why the final instar is much more resistant to baculovirus infection than the earlier instars? (SAQ 9.2 - 1) might help you formulate one).

9.3 Baculovirus genetics

9.3.1 Gene regulation

sequence of
gene
expression

Baculoviruses have a unique, temporally regulated, bi-phasic replication cycle (Figure 9.5). Use this figure to follow the description. In the first phase non-occluded viruses (NOVs) are produced; in the second, occlusion bodies (OBs) are generated. NOVs are responsible for the systemic spread of the virus between host cells whereas OBs are the infectious form which spread the virus from larvae to larvae. Gene regulation occurs at the transcriptional level. Upon infection of cells, viral genes of the immediate early (IE) class are transcribed by host factors, including RNA polymerases. Products of these genes turn on (transactivate) an array of delayed early (DE) genes, including virus coded RNA and DNA polymerase. They also turn off some host functions. Transcription of these DE genes precedes the onset of DNA replication. Late genes (L) are switched on concurrently with the onset of DNA replication and their expression is promoted by IE and DE gene products. The late genes code for structural proteins of the virus particles. A fourth class, unique to baculoviruses, represent the 'very late' (VL) genes involved in polyhedron morphogenesis (poyhedrin) and cell lysis. Genes of each of the four classes are not clustered, but randomly distributed along the genome (Figure 9.3). The regulated expression of baculovirus genes is reflected by the presence of various virus specific proteins at different times after infection (Figure 9.5).

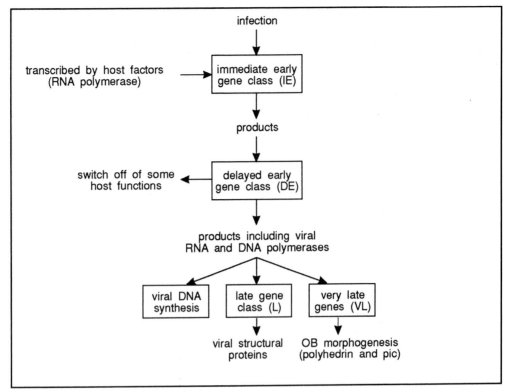

Figure 9.5 Sequence of genes expressed during viral infection (see text for details).

Regulation of baculovirus gene expression occurs at the level of transcription. With baculoviruses, the transcription products (ie mRNA) do not contain introns (sections of nucleotides that are removed during RNA processing). Some genes are differentially regulated in time and have both DE and late regulatory elements in one promoter region. Transcription is enhanced by so called enhancer sequences.

Enhancer sequences are sequences of nucleotides which increase the effectiveness of a promoter. Consider the following gene structures:

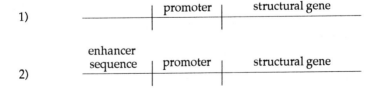

In the presence of the appropriate factors (eg RNA polymerase) the situation shown in 2) above will lead to a greater rate of transcription than that shown in 1).

The cascaded character of baculovirus gene regulation is similar to that of other large DNA viruses (eg adeno or herpesviruses). Baculoviruses have an additional, unique class of very late (VL) genes, which are hyperexpressed very late after infection and code for proteins that are involved in the final stages of virus infection and OB morphogenesis. Polyhedrin and a protein of 10 kDa (p10) are the most prominent products of genes of this class. Mutations or deletions in these very late genes do not affect NOV replication.

9.3.2 Gene structure

fine structure of baculovirus genes

The coding regions of baculovirus genes are contiguous and not usually spliced. IE and DE baculovirus genes have promoter elements similar to those of eukaryotic organisms. A CAGT sequence, where initiation of transcription occurs, and an upstream TATA box are highly conserved. One of the IE mRNAs displays splicing in the leader sequences. Transcripts of late genes are initiated from a consensus promoter element, TAAG, specific for baculoviruses. The very late genes polyhedrin and p10 are present in single copies and are hypertranscribed from a consensus promoter, AATAAGTATTTT, which includes the TAAG motif of all baculovirus late promoters. This promoter element is located at position - 49 (polyhedrin) and -70 (p10) from the translational start (ATG) of these genes (Figure 9.6). The integrity of the leader sequence upstream from the translational start is absolutely essential for high level of expression.

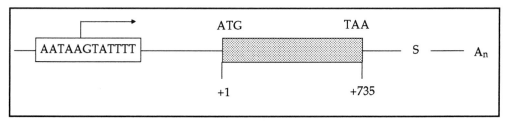

Figure 9.6 Structure of the promoter sequence of polyhedrin of AcNPV (see text for details).

∏ To help you remember these gene structures, draw block diagrams like that shown in Figure 9.6 to show the structure of immediate early, delayed early and late genes of baculoviruses. Use the description given above to help you.

Baculoviruses are similar in size and structure. The rod shaped nucleocapsids vary in length and apparently are able to accommodate DNA ranging from 80 - 200 kilobase pairs in size. Although there is limited genetic relatedness between baculoviruses in terms of overall nucleotide sequence homology, the structural organisation of genes appear to be similar. Genes with similar functions have a similar physical location on the genome. Apparently, the organisation has been conserved during evolution. In contrast, polyhedrin genes have a high degree of both nucleotide and amino acid sequence homology.

9.3.3 Gene function

genes of baculovirus with identified function

The function of a small number of baculovirus genes is known and their locations are shown on the physical map of AcNPV in Figure 9.3. Among these are two immediate early genes IE-1 and IE-N (transactivators, ie activators of other genes) the major nucleocapsid protein, the major NOV envelope glycoprotein (gp64), DNA polymerase, a basic histone like DNA binding protein, the OB envelope gene, polyhedrin and p10. Many more open reading frames (ORFs) have been detected and mapped on the genome, but their function is not yet known.

role of gene in hormone inactivation

Some genes specifically interfere with the insect metabolism. For example the ecdysone UDP glucosyl transferase (EGT) inactivates ecdysteroids by conjugating glucose. As a result the insect does not moult into the next larval instar and this allows the virus to complete its replication in the host. The EGT gene is expressed early after infection and may have been picked up from the insect host during evolution.

polyhedrin and p10 in occlusion bodies

The polyhedrin gene is the most prominent hyperexpressed gene active very late after infection. In OBs polyhedrin is arranged in a paracrystalline structure, which provides protection to occluded virus particles in the environment. p10 is located in fibrillar structures in the nucleus and cytoplasm of infected cells. These structures form a network in the cell, but a further function has not yet been assigned. Deletion of the p10 gene does not impair OB formation, but impairs lysis of cells and insect liquifaction.

SAQ 9.3

Describe how you could determine whether the high level of expression of polyhedrin, or p10, are due to a strong promoter or to the presence of multiple copies of these genes on the genome or high translational efficiency.

SAQ 9.4

Why would it be to the advantage of the virus to pick up a gene such as ecdyson UDP glucosyl transferase?

SAQ 9.5

List two molecular reasons why baculoviruses may have a limited host range.

9.4 Genetic engineering of baculoviruses expression vectors

A number of unique features promote baculoviruses as vectors for the high level expression of foreign genes in eukaryotic (insect) cells:

- Late after infection two viral proteins accumulate to very high levels, polyhedrin and p10. Their genes are hyperexpressed due to the presence of very strong promoters;

- Both proteins are involved in the morphogenesis of OBs and are not essential for the replication of infectious NOVs. The coding sequences for these proteins are therefore dispensable and can be replaced by foreign genes. The promoters are available to drive the expression of these foreign genes to high levels;

- The circular viral genome can be expanded in size and can accommodate up to 25 kbp of additional DNA, while still being properly assembled in rod shaped virus particles;

- Post translational modifications, such as glycosylation, phosphorylation, amidation etc in insect cells appear to be very similar, if not identical, to those in other, higher eukaryotic systems.

baculovirus
DNA segment
in bacterial
plasmid

cloning site
adjacent to
polyhedrin
promoter

The genetic engineering strategy of baculoviruses is based upon the allelic replacement of a baculovirus DNA segment by foreign gene elements. The baculovirus genome size (about 130 kbp) is too large to manipulate directly. Special transfer vectors are hence designed which contain, in addition to a bacterial plasmid element, a baculovirus DNA segment with a cloning site downstream from the polyhedrin promoter and 5′ and 3′ flanking sequences (Figure 9.7). The manipulations are carried out on such a plasmid and can range from introducing deletions, mutations or insertions of foreign genes.

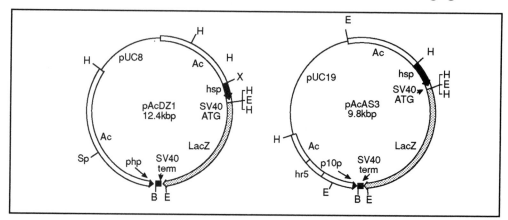

Figure 9.7 Structure of transfer vectors pAcDZ1 (polyhedrin vector) and pAcAS3 (p10 vector). E = *Eco*R1; H = *Hind*III; B = *Bam*HI; Sp = *Sph*I; hsp = heat shock promoter; php = polyhedrin promoter; p10p = p10 promoter; term = terminator; LacZ = β-galactosidase gene. The arrows indicate the direction of transcription.

For optimal expression of foreign genes it is essential to maintain the complete promoter sequence. Single or multiple unique cloning sites are engineered behind the polyhedrin promoter sequence to facilitate insertion of the foreign gene. Recombinant transfer vectors usually contain the foreign gene with its own ATG start codon and a short non translated leader sequence. The foreign genes are transferred to the baculovirus genome by homologous recombination between the transfer vector and wild type baculovirus DNA during replication in insect cells (Figure 9.8). The manipulated segment is targeted to the correct location on the viral genome by the homologous 5′ and 3′ flanking sequences.

In case of the allelic replacement of polyhedrin, recombinant viruses are usually recognised in the light microscope by the absence of OBs, or sometimes by the presence of the foreign gene (Southern blot hybridisation), or recombinant protein (immunofluorescence or immunoblot analysis). The insertion of foreign genes into the p10 locus is slightly more complicated because phenotypical markers for the absence of p10 are lacking. However, a new generation of transfer vectors is available, which facilitates the screening for recombinants by virtue of the expression of β-galactosidase as a marker (Figure 9.7).

When the recombination is successful the polyhedrin or p10 promoters drive the expression of the foreign gene to levels equivalent to the authentic polyhedrin or p10 itself, as we exemplified by the expression of β-galactosidase (Figure 9.7). This system has enormous potential for producing valuable protein products. The heterologous proteins expressed using baculovirus polyhedrin or p10 recombinants have a wide range of applications including the use as diagnostics and therapeutics in medical and veterinary practice. Hundreds of proteins from viral, bacterial, animal and plant origin

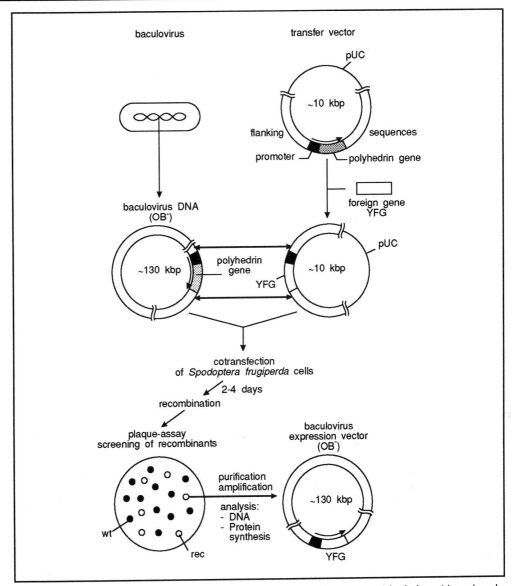

Figure 9.8 Construction of baculovirus expression vectors. The polyhedrin gene (shaded area) is replaced by 'Your favourite gene' (YFG). wt = wild type virus (OB$^+$); rec = recombinant virus (OB$^-$); kbp = kilobase pairs.

are produced via baculovirus expression vectors. The first recombinant vaccine against HIV was prepared from baculovirus-expressed gp160 subunits.

9.5 Strategies for the engineering of baculovirus insecticides

Genetic engineering of baculovirus insecticides aims at extending their host range, improving their virulence, increasing the speed of action and altering the persistence. Although host range determinants and virulence factors have not yet been discovered, improvement of insecticidal action is within reach. The ideal improved baculovirus insecticide should have a broader but defined host range and upon infection cause immediate cessation of feeding without killing the host instantaneously (Why?). It is only commercially attractive to develop baculovirus insecticides with these specifications.

9.5.1 Insertion of genes interfering with the insect metabolism

genes with insecticidal potential

A strategy for the improvement of baculovirus insecticides is the introduction of genes coding for insect toxins, hormones or metabolic enzymes. Table 9.3 lists some examples.

Class	Protein	Potential
Toxins	*Bt* toxin	+
	spider toxins	+/-
	mite toxins	+/-
	trypsin inhibitor (TI)	?
Hormones	ecdysone hormone (EH)	+
	diuretic hormone (DH)	+
	prothoracicotropic hormone (PTTH)	+
	allatotropin - promotes JH production	-
	allatostatin	+
	proctolin	+
Enzymes	juvenile hormone esterase (JHE)	+

Table 9.3 Possible candidates of insecticidal genes for introduction in baculovirus insecticides.

We will explore their modes of action later. These peptide substances interfere with the insect's metabolism and metamorphosis. High level expression via baculoviruses may improve the insecticidal properties of the viruses. Recombinant baculoviruses containing genes of this nature can be considered as natural, insect-specific biocontrol agents producing biorational compounds. The mode of action of the various gene products to be expressed via baculoviruses is outlined below.

gene product exerting effect in insect guts

Three gene products exert their insecticidal action in the gut: the toxin of *Bacillus thuringiensis* (*Bt*), trypsin inhibitor and proctolin. In chapter 8 we learnt that the spores of *Bt* contain a large proteinaceous crystal. When this crystal is ingested by insects, it is dissolved and degraded by the action of gut proteases, into toxic polypeptides (delta endotoxins). These toxins bind to gut cell receptors generating small holes in the cell membrane. This quickly destroys the regulation of ion exchange and results in paralysis of the gut and mouth parts, followed by the cessation of feeding. An array of *Bt* toxin genes into baculoviruses could enhance their insecticidal activity. An alternative strategy would be the expression by baculovirus of a trypsin inhibitor (TI), which eliminates the action of gut proteases involved in food digestion. Proctolin is a myoactive pentapeptide increasing the frequency of hindgut contractions. It is a neurotransmitter, but it has a severe effect on the gut when administered with the food.

consequences of over production of regulatory molecules in insect larvae

Some morphogenetic regulatory gene proteins act in the haemolymph. Juvenile hormone esterase (JHE) regulates the concentration of juvenile hormone (JH). Decrease of JH level in the haemolymph through JHE activity leads to precocious (early) metamorphosis of the young larvae, which is in turn preceded by cessation of feeding. The same developmental effect is observed with baculoviruses lacking the EGT gene.

Diuretic hormone (DH) regulates the water balance and possibly blood pressure in insects. High levels of DH cause cessation of feeding, excretion of water, desiccation of the insect and early mortality. Ecdysone hormone (EH) initiates ecdysis, the process leading to the shedding of the old skin. Overproduction of this hormone would enhance this process and cause premature death. Prothoracicotropic hormone (PTTH) is involved in triggering the moulting process. PTTH is synthesised in the brain and stimulates synthesis and release of ecdysteroids by the prothoracic gland. JHE, PTTH and EH induce premature moulting and, together with JH, affect metamorphosis and ecdysis.

JH is a terpenoid and plays a vital role in the control of insect morphogenesis and reproduction. It is produced in cells of the *corpora allata* in the insect brain. High levels of JH maintain the larval stage. The release of JH in the haemolymph is regulated by the activity of the *corpora allata* which produce allatotropins (+) and allatostatins (-). Allatostatin is a good candidate to be expressed by baculoviruses. Upon infection insects will stop feeding and start to moult.

Π   Make a list of these factors which could be used to increase the insecticidal activity of baculoviruses. Then imagine you are developing such viral strains, rank them in order you attempt to produce them. As you rank them, decide what kind of criteria you are using.

We believe you are likely to be thinking of: the likely efficiency of the viruses with the new gene inserts as insecticides and how difficult it will be to isolate the relevant genes to be inserted (eg what source to be used, are there suitable probes for such genes etc).

The protein sequence of most of these hormones and enzymes is known and in some instances the authentic gene has been isolated. From the protein sequence synthetic genes can be made and inserted into baculoviruses. Signal sequences can be added to enhance secretion of proteins. Overproduction of toxins, hormones or regulatory enzymes will invariably upset the insects metabolism, which may result in cessation of feeding, premature moults, desiccation or premature death. It is possible that the insect counteracts this by producing compensatory products in order to restore homeostasis. Eventually insects will die from the virosis.

use of antisense   A completely different strategy may be the antisense approach. We learnt in earlier chapters that expression of bacterial and eukaryotic genes can be blocked by the production of antisense RNA. A copy of the coding sequence of the target gene is engineered in the opposite orientation downstream from a strong promoter. The RNA thus has an antisense character and will form double stranded RNA with the sense RNA, which can as a consequence not be translated into a protein. This strategy has been successfully used to alter flower colouring and to suppress virus replication in plants. It can be envisaged to include important insect genes coding for metabolic or regulatory enzymes, or hormones, in the antisense orientation under the control of a baculovirus promoter. As a result the expression of this antisense gene would immediately block essential host functions upon infection of the insect.

9.5.2 Deletion mutants

EGT gene   As part of the infection strategy of viruses some genes are involved in re-directing the host cell machinery towards optimal production of the virus. It would be to the disadvantage of a baculovirus to kill the insect immediately. This would reduce its chance for survival as only small amounts of progeny virus will then be produced. The presence of the ecdysone UDP glucosyl transferase (EGT) gene in baculoviruses is essential to maintain the larval stage and to prevent another moult. Baculoviruses lacking the EGT gene induce precocious metamorphosis of larvae, which is in turn preceded by cessation of feeding.

potential of deleting envelope genes

Virus host interaction is the result of a long co-evolution history. Deletion of viral genes in general will not be to the selective advantage of the virus as they would affect the virulence, yield or alter persistence. Deletion of the polyhedrin gene results in the inability of progeny virus to infect insects by the oral route, whereas deletion of other genes impairs virus production or persistence.

9.5.3 Choice of promoters and sites of insertion

Ideal sites for insertion of foreign genes are those areas of the genome which are not essential for virus replication in insects or insect cells. Most genes involved in the late stages of virus morphogenesis (eg polyhedrin, polyhedron envelope protein), are usually dispensable for virus replication, but essential for OB morphogenesis.

co-infection - a strategy for overcoming loss of polyhedrin genes

To obtain recombinants with increased insecticidal activity, high levels of the insecticidal protein in the gut, haemolymph or other tissues are usually required. The polyhedrin and p10 promoters meet most of these requirements. Insertion of insecticidal genes at the polyhedrin locus has, however, the disadvantage that the polyhedrin gene is lost and that virus transmission among insects is no longer possible as OBs are no longer produced. To alleviate this problem recombinant viruses are co-occluded with wild type (wt) baculoviruses in an OB. Such OBs thus contain viruses with wt and recombinant genotypes. The wt baculovirus provides the polyhedrin for the formation of OB and complements the OB negative phenotype of the recombinants. Upon infection of insects co-occluded recombinants can exert their insecticidal effect (Figure 9.9).

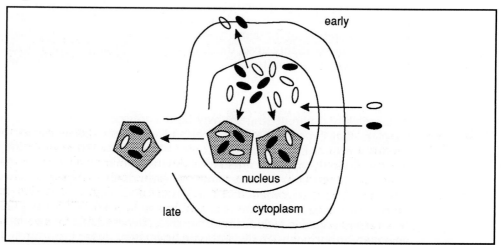

Figure 9.9 Co-occlusion of wild type (open rods) and recombinant (closed rods) baculoviruses during co-infection of an insect cell. Note that in the late stages OBs containing both types of viruses are produced.

Alternatively, recombinants are constructed where the polyhedrin gene is maintained, but relocated into another site on the viral genome, proximal to its authentic location. The original polyhedrin promoter is now able to drive the expression of the insecticidal gene. The transfer vectors hence contain both the polyhedrin gene in its original location and the foreign gene under the control of a duplicated polyhedrin promoter (Figure 9.7).

This approach has the added advantage that recombinants can be conveniently retrieved after co-transfection with OB negative viruses by the presence of OBs.

The locus of the p10 gene is ideal for the purpose of inserting insecticidal genes. It is dispensable for virus replication and does not interfere with OB formation. Its strong promoter can drive the expression of insecticidal genes, such as insect toxins, hormones and enzymes.

site of production of insecticide and site of action timing of insecticide expression

In most baculoviruses polyhedrin and p10 are not expressed in the gut, whereas some of the insecticidal proteins, eg *Bt* toxin, proctolin, act on gut cells. In addition, the promoters of these two genes are expressed late after infection (Figure 9.5) which may result in delayed insecticidal action. This implies that in order to provoke an immediate insecticidal response immediate early, delayed early or host promoters are more suitable than the major late polyhedrin or p10 promoters. The insecticidal proteins would be expressed much earlier, although at a reduced level as these promoters are weaker. However, it may not always be an advantage to kill the insect too quickly, as progeny OBs cannot be produced.

SAQ 9.6

1) Why would host promoters be equivalent to the IE promoter of baculoviruses for driving the expression of foreign genes? What would be the disadvantages to using these promoters for driving insecticidal genes?

2) What would be the effect of trypsin inhibitor expression by baculoviruses?

3) Why might it be expected that recombinant baculoviruses producing *Bacillus thuringiensis* toxin may not be an effective insecticide?

4) Explain why baculoviruses lacking the gene for ecdysone UDP glucosyl transferase (EGT) cause larvae to moult early.

5) Why would the expression of an allatotropism gene by baculovirus be ineffective to control insects? (Think about the activity of allatotropin - see Table 9.3).

9.5.4 Safety considerations

importance of specificity

The specificity of baculoviruses provides an inherent safety to their use in the environment. AcNPV has the widest host range known amongst these viruses and infects 58 insect species. Usually, baculoviruses infect a single or a few related insect species. They do not adversely affect non target hosts including vertebrates, plants and beneficial invertebrates and therefore impose no ecological hazard on the environment unless beneficial insects are also within the host range. In the case of genetically engineered baculoviruses for insect control it is desirable to have recombinants with increased insecticidal activity, but with limited ability to survive or persist in the field. They should preferably have reduced biological fitness as compared to the wt baculovirus.

reducing virus spread

To achieve these goals, genes can be deleted from a baculovirus following a similar strategy as with the insertion of genes. These alterations may result in reduced virus persistence or progeny yield, which greatly affect their epizootiological potential. For example, the gene for ecdysone UDP glycosyl transferase (EGT) from the virus prevents the activation of ecdysone in the insect upon infection and allows the insect to increase in size and to produce large amounts of OBs. When this gene is deleted insects moult normally, but produce less OBs. Deletion of the p10 gene affects cell and insect lysis, limiting the dissemination of the virus in the environment, whereas deletion of the OB envelope gene results in reduced persistence. Allelic replacement of the p10 gene with an insecticidal gene therefore has a double impact.

When the polyhedrin gene is replaced by an insecticidal gene the recombinants also are not able to form OBs and to infect insects efficiently. In this form virus particles are unable to persist in the environment for a long period of time (ie more than a day or two). Yet, in order to disseminate these recombinants they can be co-occluded into an OB of a wt baculovirus during a co-infection. Upon release in the environment the recombinants have a selective disadvantage over wt baculoviruses, since only OBs are infectious.

A major concern with the use of baculovirus recombinants is the possibility of transfer of the insecticidal gene to other, distantly related baculoviruses or to non target hosts. Although baculoviruses have a similar genomic organisation, the nucleotide sequence homology is low and illegitimate recombination has never been observed. Transfer of baculovirus genes to non target insects has not yet been observed. The presence of insect transposon like elements in baculoviruses presumably results in genetic flow from insects to viruses, rather than in the reverse direction.

SAQ 9.7	List two main concerns over the release of genetically manipulated baculoviruses.

9.6 Production of (recombinant) baculoviruses

in vitro and in vivo cultivation

Efficient production of baculovirus (recombinants) requires a suitable and economic production process. Baculoviruses can be produced in insect cell cultures as well as in insect larvae. There are limitations and advantages to both systems. It is essential in both cases that OBs are produced to allow infection of insects.

9.6.1 Production in insects

raising insect larvae not automatic

Insect larvae serve as an excellent 'bioreactor' for baculovirus production. Approximately $1\text{-}2 \times 10^9$ OBs are produced per larva, which amounts to 25-30% of the larval dry body weight. Larvae are easy to rear and can be obtained at relatively low costs. However, each baculovirus requires a specific insect host and since insects are sometimes cannibalistic, they cannot necessarily be reared in an automatic way. This raises labour costs and makes the commercial production of baculoviruses expensive.

Since OBs are used for infection of insects, recombinants should be of the co-occluded type, or have maintained the polyhedrin gene either in its authentic location (eg using the p10 locus) or relocated to another part of the genome. Recombinants with insecticidal activity can only be produced in insects when this activity does not immediately lead to mortality. Otherwise, insect cells are the only viable alternative for production of recombinants.

9.6.2 Insect cells

Advantages of insect cell cultures for the production of baculoviruses are:

- convenient quality control;

- minimal contamination;

- convenient down stream processing;

- less cruelty.

Disadvantages are:

- the costs of insect cell culture media;

- the absence of high volume bioreactor systems.

suitable cell type

There are many insect cell lines, usually derived from young larvae or from ovaries of adult females. Only a few replicate baculovirus in a fully permissive fashion and have been characterised in some detail. The most appropriate virus cell system to date consists of cell lines of the fall armyworm *Spodoptera frugiperda* and the cabbage looper *Trichoplusia ni*. These cells are relatively easy to grow in large volumes, the major problems being sufficient oxygenation in the bioreactor and the sensitivity of the cells to hydrodynamic forces.

cultivation in spinner or tissue culture flask

For production of baculoviruses small scale tissue culture, flasks or spinner cultures (up to 1 litre) are used to propagate cells and viruses. Here, oxygen is provided through surface aeration and the cells are kept in suspension by conventional stirring of the cell suspension. Cell densities of $1-2 \times 10^6$ cells and an OB production of 2×10^7 per ml can be obtained. In larger volumes surface aeration is insufficient to meet the oxygen demand for optimal cell growth and virus production. Introduction of oxygen through permeable tubing (sparging) has been successful at medium size volumes (up to 10 L), but this is impractical in larger volumes.

airlift reactor

The logical alternative is the use of airlift bioreactors. Rising air bubbles keep the cells in suspension and provide the oxygen required. The hydrodynamic forces cause irreversible damage to the cells. This effect is reduced in large airlift bioreactors by using special reactor designs and medium. Reactors up to 1000 L can be used. If the reader requires further information regarding these methods of cell cultivation, we recommend the BIOTOL texts relating to *in vitro* cultivation of cells and bioreactors.

9.7 Application

inoculative approach

There are two major strategies for the release of baculoviruses in the field. The first strategy is the inoculative release of a baculovirus in the epicentre of the insect pest. This virus will cause an outbreak of the viral disease and provide natural long term control of the pest population. The disease will remain endemic in the area treated. However, this strategy can only be applied in stable ecosystems (forests, grasslands) where some damage can be tolerated and high densities of insects occur.

SAQ 9.8

1) Given that each larva weighs a few grams, in which order of magnitude (kg, g, mg, µg or pg) can recombinant proteins be produced per larva assuming that the level of recombinant protein is similar to polyhedrin?

2) Given that about 10^{11} OBs are required per hectare and that about 50 OBs are produced per cell in culture, what size of reactor would be needed to produce sufficient OBs to treat one hectare of crops?

3) Would it be possible to infect insects other than with OBs and how would this be achieved?

4) Can you explain why young (neonate) larvae or adult female ovaries are used to develop cell lines? (Think about what will normally happen to the cells from these sources).

inundative approach

In most cases, however, the damage threshold is low and high population densities of insects, required to establish and maintain an endemic disease cannot be tolerated (cultivated arable crops, greenhouses). In addition, intense agro-ecosystems with frequent harvest reduce the possibility of establishing an endemic disease. Under these circumstances inundative releases at frequent intervals are required for sufficient control. In order to obtain adequate short term control about 10^{11} OBs per hectare are required, equivalent to approximately 50 infected larvae.

monitoring viral survival and gene transfer

Application of recombinant baculoviruses at a large scale has not yet been performed. Only limited releases have been carried out, mainly for risk assessment studies. Monitoring of the behaviour of recombinants in the environment is an important aspect of these studies. Marker genes, such as β-galactosidase, are significant elements in the study of virus ecology and possible gene transfer to other organisms and viruses.

SAQ 9.9

What strategy should be considered to control insects in greenhouses with baculoviruses?

9.8 Conclusions

In this chapter, the biology and molecular genetics of baculoviruses have been reviewed. The possibilities and strategies of genetic engineering to improve the insecticidal activity of these pathogens have been outlined. The detailed knowledge about baculovirus gene structure, function and regulation allows the manipulation of the viral genome and the development of the baculovirus expression vector system. This technology has been applied and tailored for the combat of insect pests. The construction of baculovirus recombinants which cause immediate cessation of feeding is now within reach. In addition, by deletion mutagenesis these recombinants can be made less persistent in the environment which reduces the environmental risks associated with deliberate release. The recombinants can be produced in insects, when the insecticidal action does not cause immediate mortality. Otherwise, they can be produced in cell culture. Development of additional strategies for improvement of baculoviruses requires a more detailed understanding of insect biochemistry and physiology.

Summary and objectives

This chapter has explained how genetic engineering can be used to produce more effective biological control agents against pest insects. The discussion focused onto baculoviruses but many of strategies described have wider applications.

Now that you have completed this chapter you should be able to:

- describe the structure and occurrence of baculoviruses in terms of their chemical composition and their host ranges;

- explain the difference between viruses in occlusion bodies and naked viruses;

- give reasons why the promoters of polyhedrin and p10 are excellent candidates for the expression of insecticidal genes;

- explain the evolutionary and biological significance of polyhedrin and ecdysone UDP glucosyl transferase inheritance by baculoviruses;

- list strategies for engineering baculoviruses to make then more effective as insecticides;

- list the major environmental considerations in identifying the risks involved in the release of genetically engineered baculoviruses;

- compare the advantages and disadvantages of using larvae and cell cultures for producing baculoviruses;

- identify circumstances when inoculative and inundative approaches to using baculovirus insecticides should be used.

10 Diagnostics in plant biotechnology

10.1 The use of diagnostics in plant protection   244

10.2 Two basic principles for the detection of plant pathogens   245

10.3 Current detection methods in plant protection   245

10.4 Biotechnology and the detection of plant pathogens   253

Summary and objectives   255

10 Diagnostics in plant biotechnology

10.1 The use of diagnostics in plant protection

In plant pathology, as in other pathology fields, diagnostics are used to ascertain the nature of an agent that is causing a disease. However, in many cases diagnostics are used to ascertain if a pathogen is present in plant material whether or not the plant is showing the symptoms of a disease. In the latter cases one should speak of detection methods rather than of methods and means for diagnosis.

Diagnostics are used by Plant Protection Services to prevent import of pathogens that are of potential risk for a particular country. Frequently disease agents are involved that can cause serious losses in the agricultural industry, either by causing direct damage (eg losses in yield) or by causing indirect damage, (less export of (propagating) material if the disease is present in the exporting country).

certification of
disease free
status

In the Netherlands, detection methods are also used for agricultural products by the General Inspection Services. A lot of propagating material is sold with a certificate that is issued after testing for quality with respect to trueness to type and for the health status of the material. Other nationalities offer analogous schemes.

Many tests for the presence of pathogens are also performed by companies that want to multiply plants by the types of rapid biotechnological technique described in earlier chapters. It is essential for these activities to start with absolutely healthy material. In breeding programs it is important to make sure that the starting material is free from pathogens especially from viroids and viruses that may be transmitted by seed and pollen. It is known that these disease agents can be transmitted readily in this way.

Detection of pathogens is important too in ecological studies, eg to study the effect of biological control of fungi and bacteria by antagonists. Under such circumstances it is necessary to be able to ascertain population densities by quick, reliable methods that can be applied on a large scale.

One can imagine that different diagnostic methods may lead to different standards. To prevent the spread of disease agents or to make sure that mother plants to be used for mass multiplication are healthy, one needs highly reliable methods. It is not an absolute prerequisite for the tests to be cheap but the price of the test is nevertheless a very important aspect. If the tests are expensive, then this will add significantly to the cost of the product.

This final chapter examines the principles of disease detection methods and explores how biotechnology is contributing to these important areas.

10.2 Two basic principles for the detection of plant pathogens

bio-assay of pathogen

Detection methods can be based either on the relationship between a pathogen and host plants or on the intrinsic properties of the pathogen. In the first case, one tries to transmit, mechanically or by means of a natural vector, an inoculum made from the diseased parts of the affected plant to a set of plants, from which the reaction to several pathogens is well-known.

sensitivity

overlap of symptoms

time consuming

expensive

non-automated

This biological way to detect pathogens is, in principle, very sensitive. For instance in the case of plant viruses one particle of potato virus Y, when presented to an optimal infection site, is able to infect a potato or a tobacco plant systemically, producing typical symptoms. In such cases one can detect 83×10^{-18} gram of virus. However, many pathogens produce very weak symptoms or none at all and symptoms are not always unique. Furthermore the use of test plants is time-consuming; the plants have to be cultured and depending on the combination of plant and pathogen, it may take from 2 days up to several weeks for the symptoms to develop. The use of test plants is expensive, especially when the plants have to be grown in the winter season at the expense of additional heating and illumination. Reading symptoms requires skills of a specialist and this prevents these detection methods from being automated.

∏ It may be useful to make yourself a table listing the advantages and disadvantages of using bio-assays for detecting the presence of disease agents, use the headings **Advantages, Disadvantages**.

assay of pathogens

Methods based on intrinsic properties of the disease agent use such properties such as shape and dimensions; type, number and molecular mass of nucleic acids and proteins; tertiary structure; the presence of specific antigenic sites). They can in general can be automated and therefore are very attractive for large-scale screening programmes.

10.3 Current detection methods in plant protection

In this chapter we will concentrate on detection methods based on intrinsic properties. Furthermore we have limited ourselves to plant pathogenic viroids, viruses, bacteria and fungi.

polyclonal

monoclonal

We will assume that the reader is familiar with the immune-response system of vertebrate animals, the production of polyclonal and monoclonal antibodies, and the fact that antibodies can also be used in *in vitro* to specifically detect the antigen that was used to induce their formation. If the reader is unfamiliar with these we recommend the BIOTOL test "Technological Application of Immunochemicals". The essential point to remember is that antibodies bind specifically with target structures (antigens). We make use of this feature whether or not the antibodies are polyclonal (mixtures) or of a single type (monoclonal). Also the production of probes for molecular hybridisation and their use to detect specific sequences of bases in nucleic acid molecules will not be explained here. We have, however, discussed this aspect in some detail in Chapter 7.

10.3.1 Detection of viroids

viroid

naked RNA

A viroid is a small plant pathogenic agent that consists of a circular single-stranded RNA with a high degree of internal base-pairing. Potato spindle tuber viroid (PSTV) is internationally considered to be of quarantine importance. It is not present in the

Netherlands and therefore a reliable method was needed to test potato material imported for breeding purposes. Large numbers of PSTV strains produce very weak symptoms in potato and other host plants and therefore it is important to find a biochemical or biophysical test to be able to ascertain the presence of these agents.

∏ Viroids contain no proteins. Is it possible to develop a specific antibody against these to be used in serological tests? (We will give you an opportunity to check you answer later).

Polyacrylamide-gel electrophoresis

polyacrylamide-gel electrophoresis

The first method used to detect viroids was based on the detection of the specific circular single-stranded RNA molecules. Nucleic acids were purified from plants to be tested and the fraction of small molecules (soluble in 2 mol I^{-1} LiCl) was analysed on a 5% polyacrylamide-gel. The nucleic acids were stained with toluidine blue. The method was not very sensitive and the staining was not very powerful. Therefore Morris and Smith developed an alternative in which the initial nucleic-acid extract was inoculated onto a tomato cultivar (Rutgers). The plant was then grown for several weeks at 28°C under continuous light. During this period the viroid multiplied in the plant. Subsequently another nucleic acid extract was made from the tomato plant and analysed on gel. This made it possible to use combined samples for the initial nucleic acid extraction. One diseased plant amid 199 healthy ones could be detected. However a lot of confusion still existed in the data because of the weak toluidine blue staining method.

Molecular hybridisation

molecular hybridisation

nick translation

In 1981 Owens and Diener published good results with the use of molecular hybridisation. For this, a full length cDNA against PSTV was constructed and cloned into the plasmid pBR322. This plasmid contains ampicillin and tetracyclin resistance genes. Insertion of foreign DNA into one of these genes produces a plasmid carrying only a single functional drug resistant gene. This enabled selection of hosts carrying the plasmid containing the foreign DNA. The principles of this type of vector isolation and use were described in Chapter 1.

Nick translation involves the following steps.

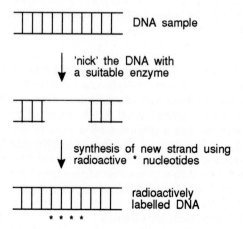

The cDNA derived from PSTV was labelled with ^{32}P using nick-translation.

This cDNA was used as a probe in hybridisation studies with nucleic acid extracts from tomato and potato plants spotted on nitrocellulose filters. The extraction procedure was kept simple using 0.5 ml of sample obtained from 1 g of leaf material. Concentrating the sample even more increased the sample-preparation time dramatically. In this way a spot in the autoradiogram could be found if at least 125 pg (pg = 10^{-12}g) of viroid amid other nucleic acids was present on the filter. The amount of sample that could be applied was 3 microliter (μl).

SAQ 10.1

What concentration of viroid in a plant can be detected in terms of ng (10^{-9}g) per gram of leaf using this procedure?

A large disadvantage of the method was the use of the radioactive label and therefore the method never became popular for large-scale routine testing. Nowadays very good non-radioactive labels are available, but the molecular hybridisation method for the detection of viroids is still not used because easier methods have been developed.

Bi-directional polyacrylamide-gel electrophoresis

To be able to use a more powerful stain after polyacrylamide-gel electrophoresis than toluidine blue, it was necessary to reduce the amount of background material in the region where the viroid had to be detected. This is not possible in a single electrophoretic separation step. The use of more sensitive stains (eg $AgNO_3$) would result in a more-or-less evenly stained gel. This was overcome in the following way. After electrophoresis under native conditions the region in which the viroid is present is cut out. Suitable markers (ie standard samples of viral RNA) are run in parallel lanes in order to identify the correct regions. This part of the gel contains single stranded RNA which co-migrated with the largely internally based-paired viroid RNA. The region with the viroid is brought to the bottom of a new gel and subjected to electrophoresis under denaturing conditions. Under those conditions (high temperatures or low salt concentrations), the highly internally base-paired viroid molecule will open up and the resulting larger molecule will move more slowly in electrophoresis than the contaminating nucleic acids that were not affected by the denaturing conditions. After the electrophoresis the viroid band can be found in a region of the gel where no other material is present and then a powerful stain like $AgNO_3$ can be used successfully to visualise it. Later on this method was modified in such a way that the first run was continued till the viroid band was at the end of the gel. Then the buffer was changed for a buffer solution at 70°C, the polarity of the electric field was changed and the gel electrophoresed for a second period.

sensitivity of bi-directional electrophoresis

This method enables the detection of about 5 ng of viroid in a band in the gel. In practice the lowest concentration that can be detected in plant material is 10 ng per gram of leaves. This also means that this so-called bi-directional electrophoresis is as sensitive as molecular hybridization.

sensitivity and combination of samples

In conclusion we would claim that currently bi-directional electrophoresis combined with $AgNO_3$ is the best method to detect viroids. The method has the same sensitivity as, but is faster than, molecular hybridisation. The method can be used for all viroids without changes in the procedure. With molecular hybridisation one needs a specific cDNA probe for each viroid. Furthermore bi-directional electrophoresis is based on two criteria: a staining at a specific place in the gel where the viroid is brought by two electrophoretic separations. The method is sensitive enough to allow combined samples to be tested. It can detect 10 ng of viroid per gram of leaves. Note that in the literature viroid concentrations of 1200 - 1900 ng per gram of leaves have been reported for viroid infected chrysanthemums and potato tissue. In the Netherlands for example the General Inspection Service for Ornamental Plants is combining 20 samples while testing chrysanthemums for chrysanthemum stunt viroid. In other words only 1/20 of the number of tests need to be done compared to tests using single samples.

∏ You will probably find it useful to draw yourself a summary chart of the advantages and disadvantages of the various techniques we have described for detecting viroid for each method. Use the headings **Advantages/Disadvantages.**

SAQ 10.2

Identify the viroid RNA band in the final gel drawn below.

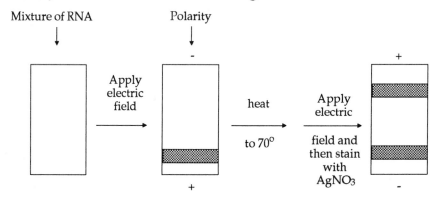

10.3.2 Detection of viruses

use of Ab against coat protein as diagnostic agents

Most plants viruses are rather simple agents that consist of 1-4 nucleic acid molecules that are surrounded by a protein coat, (capsid) that is built from one or two types of subunits. About 90% of the plant viruses have a single-stranded RNA genome; 4% have a double-stranded RNA genome and 6% have a DNA genome. Both the protein coat and the nucleic acid can be used for detection purposes. The coat protein in most cases can be used as a very efficient antigen to raise antibodies for serological detection methods. The nucleic acid part of plant viruses can function, for example, as a target in molecular hybridisation tests. Nucleic acids give rise to unspecific antibodies and therefore can not be used in serology. (Look back to the ITA at the beginning of 10.3.1, did you get the right answer?).

Serological methods

ELISA

Serology was introduced for the detection of viruses in flower bulbs as early as 1938. Serology is still very popular because over the years the methods used to purify antigens have improved tremendously and also many new methods for using antibodies, like the enzyme-linked immunosorbent assay (ELISA), have emerged. These systems use an enzyme attached to the antibody to detect if the antibody has bound to its antigen.

monoclonal antibodies

One can use two types of antibodies. Polyclonal antibodies are produced by injecting a vertebrate animal with an antigen preparation and collecting the blood serum. Monoclonal antibodies are produced by fusing B cells from the spleen of an immunized animal with cancer cells (a process analogous to the somatic hybridisation of plant cells described earlier. In this case we attempt to combine the antibody producing features of the B cells with the growth characteristics of the cancer cells). Each fusion hybrid (hybridoma) will produce specific antibodies. Large quantities of monoclonal antibodies can be made by growing large numbers of hybridoma cells and harvesting of the antibodies from the culture fluid.

specificity of antibodies

For routine tests the polyclonal antibodies are preferred over monoclonal antibodies because they generally have a wider range of activity towards different strains of a pathogen. It should be mentioned however that monoclonal antibodies have been described that have a very wide reaction spectrum. Jordan from Beltsville, USA, for instance produced a monoclonal antibody that reacts with all known potyviruses, a group of viruses with flexuous rod-shaped particles.

latex- agglutination test

The classical serological tests, like the precipitation and the agglutination tests, could only be used for high virus concentrations. Modern methods, like the ELISA and the latex-agglutination test, are much more sensitive. The latex agglutination test involves coating latex beads with the antigen and cross-binding (agglutination) the beads with antibodies. We can stylise this as:

latex beads   antigen   coated latex beads

add antibodies

agglutination of beads

In the absence of antigen, the beads will not agglutinate. Because these tests can easily be mechanised and automated it is used a lot for large-scale testing. The assay can be used not only for leaf samples but also for seeds, bulbs, tubers and for bark material.

In The Netherlands about 10×10^6 ELISA tests were performed in 1990 (5.5×10^6 samples of potato material and 4.5×10^6 samples of ornamental plants). ELISA assays are being greatly extended to cover even more plant viruses. Internationally these now number over 100.

Molecular hybridisation

use of cDNA probes

Because most of the plant viruses have an RNA genome one has to prepare cDNA first to be able to produce cDNA or cRNA probes. The cDNA can easily be made from the genomic RNA by using the reverse-transcriptase reaction. When it is cloned in a suitable vector it then can be used to produce cDNA and cRNA for detection purposes.

Detection of plant viruses by molecular hybridisation is a powerful tool to detect small amounts of virus or viral RNA for fundamental research purposes, but it has never been used in large-scale routine testing. The reason for this is that the molecular hybridisation has about the same sensitivity as ELISA, while the application of the molecular hybridisation is far more complicated, it takes more time and it can not be automated as easily as ELISA.

10.3.3 Detection of bacteria

problem of overgrowth by saprophytes

The detection of plant pathogenic bacteria is largely hampered by the fact that in 1 ml of plant extract the target bacterium is usually amid a million or more other (mostly saprophytic) bacteria. These bacteria sometimes interfere in isolation tests, because they inhibit the growth of the target bacterium and/or overgrow it. In serological tests they may interfere because they show cross-reactivity with the antibodies used. The ratio of saprophytes/target bacterium (S/T ratio) is very important for the reliability of the test. The higher the ratio the lower the reliability.

Isolation and identification

selective media

chemical identification

Detection of bacteria by isolation methods is very sensitive if a medium can be defined that is specific for the target bacterium. The bacterium can be grown into a colony and then be identified by morphological and biochemical characteristics, with more or less 100% reliability. One of the newest methods for identification of bacteria is that of fatty-acid analysis of bacteria that are grown under standard conditions, using HPLC and a computer program that compares the elution pattern obtained with those of well-characterised bacteria (ie chemical taxonomic methods).

A serious problem for the isolation method is that the number of bacterial colonies that can be grown on the surface of a medium is limited. In a Petri-dish with a diameter of 10 cm about 150 separate colonies can be grown. If that number is exceeded, there will be problems from saprophytes overgrowing the target bacterium.

Another limiting factor for detection based on isolation is that for many bacteria no specific media are known. Efforts are being made to develop systems that will help to establish the optimal medium composition for plant pathogenic bacteria.

Because the isolation methods can not be automated they can not be used for large-scale testing. Furthermore the reliability of the method depends to a large extent on the S/T ratio.

Serological methods

∏ As you read this section, have a piece of paper ready to record the advantages and limitations of these methods. Again divide the paper into two columns one headed advantages, the other disadvantages.

Antibodies have been raised against whole bacterial cells, cell-wall components (eg lipopolysaccharides and proteins) and to specific enzymes (eg the pectate lyases of

Erwinia chrysanthemi, the causal agent of stem rot in potato). Antibodies to pectate lyases have proved to be very specific. Antibodies against cell-wall components in general are less specific and show cross-reactivity among different bacteria.

some lack of correlation between ELISA results and symptoms

In ELISA only soluble antigens of bacteria can be detected. The cells themselves are washed away during the washing steps used in ELISA. Experience has shown that ELISA results with antibodies raised against whole cells of *E. chrysanthemi* (a disease organisms of chrysanthemums) show insufficient correlation with the appearance of symptoms in the field during the next growing season. Both false positives and false negatives occur.

Serological methods to detect plant pathogens are not dependent on the S/T ratio if the antibodies used are specific and they can readily be automated. Disadvantages of these methods are the relatively low sensitivity (10^4 cells/ml for ELISA) and the fact that antibodies with a high specificity are not always available. This leads to cross-reaction with saprophytes. It is also impossible to discriminate between dead and living bacteria.

immuno-isolation

Antibodies may be used to concentrate bacteria from a mixture by immuno-isolation. The antibodies are bound to a solid phase and allowed to react with the mixture of bacteria. The target bacterium will be trapped and the contaminating bacteria can be removed by gentle washing. A new way to use antibodies for selective concentration is by binding the bacteria to magnetic beads through antibodies. The magnetic beads together with the specifically trapped bacteria can be recovered by using a strong magnet!

Immunofluorescence colony-staining

combination of immuno-isolation selective culture and immunological staining

To combine the advantages of the isolation, a serological technique has been developed in which immuno-isolation, the growing of single cells into colonies and the identification of the bacteria by serology (and if necessary further morphological and biochemical characterisation) are combined. This has become known as the immunofluorescence colony-staining method. The bacteria are selectively isolated using specific antibodies fixed to a solid phase and then mixed thoroughly with a selective agar medium. Embedded in the agar, many small colonies will be formed, there is less interference between individual colonies as compared to growing colonies on top of a plate. In this way 100-1000 or more individual colonies per plate can be obtained. However, because of their restricted growth these colonies can not be characterised by morphology. So, the agar film containing the colonies is dried and incubated with an antibody coupled to a fluorescent dye. The positively reacting colonies can then be scored with a UV microscope.

The sensitivity of the immunofluorescence colony-staining method is about 10^2 cells/ml. The reliability of the method depends mainly on the selectivity of the culture media and on the specificity of the antibodies used for immuno-isolation and for detecting the colonies by immunofluorescence.

Molecular hybridisation

Specific DNA probes can relatively easily be cloned from the DNA of plant pathogenic bacteria. However, the method is not very sensitive if one uses a probe that is specific and binds with a gene that is only present in a single copy in each bacterial cell. It has been calculated that the sensitivity is between 10^5 and 10^8 cells/ml. This has been supported experimentally.

We can conclude that in the immunofluorescence colony-straining method many advantages of the different methods to detect bacteria are combined and this method has the best prospects for the future. Research in the future will concentrate on improvement of the selective media and on ways to automate the procedures (eg the fluorescencing colonies can be analysed automatically with a computer-assisted image analyser).

SAQ 10.3	Explain why ELISA rather than molecular hybridisation has become the method or choice for detecting viruses.

10.3.4 Detection of fungi

identification based on morphology and biochemistry

The most common method used to detect and to identify fungi is to isolate and culture them on a medium and then characterise them by morphological and biochemical techniques. However, for many plant pathogenic fungi quicker methods with the same reliability are needed.

Recently monoclonal antibodies against fungal plant pathogens have gained increasing attention. Most of the work has been aimed at differences in cell membrane or cell-wall components of the mycelium and the spores, but intra-cellular components are also used as antigens. The specificity of the antibodies obtained varies greatly. They range from genus-specific to isolate-specific. Electrophoretic methods to detect specific fungal components are also being developed. Both these areas are undergoing rapid development.

10.3.5 Conclusions

importance of signal/noise ratio

The limitations of the described current methods to detect plant pathogens can be summarised by saying that the signal/noise ratios of the tests are troublesome. A bad signal/noise ratio can be caused either by a too weak specific signal or by a too high background. A low specific signal is very often the result of the need to detect very low amounts of material. For example, with bacteria it is desirable to detect a single viable bacterium because such a cell could give rise to the infection of the whole plant when it finds a favourable infection site. The same holds for single virus particles.

increase signal

Many attempts to increase the reliability of detection methods have been aimed at increasing the amount of material that has to be detected. We have already mentioned the inoculation of viroid extracts on tomato cv. Rutgers. Similarly, the culturing of the target bacterium in the immunofluorescence colony-staining method is aimed at increasing the amount of bacteria in order to facilitate their detection.

decrease in noise

Another way to increase the signal/noise ratio is to decrease the background problems. In ELISA this can be done in two ways. Background due to cross-reactions can be overcome by producing more specific antisera, either by injecting extremely pure antigens or by producing monoclonal antibodies. Non-specific binding of antibodies can be overcome by using additives like polyvinylpyrrolidone, skimmed milk powder, etc.

Background caused by the abundant presence of saprophytes in the detection of bacteria can be overcome by using more specific media in the isolation procedures.

The use of bi-directional electrophoresis for the detection of viroids is also an elegant way of solving background problems, making use of the specific tertiary-structure properties of viroids.

10.4 Biotechnology and the detection of plant pathogens

For some of the problems mentioned above biotechnology can help to find an answer. For example, by producing monoclonal antibodies one can make highly specific antibodies and with the 'polymerase chain-reaction', we described in an earlier chapter the amount of material to be detected can be enhanced.

10.4.1 Monoclonal antibodies

problem with selecting appropriate hybridomas

Monoclonal antibodies have greatly contributed in many fields of research. However, the contribution to the development of specific antibodies against plant pathogens has been limited. The main reason is that the production of monoclonal antibodies was often tried in cases where the antigen could not be optimally purified. The result was that the animals that were immunised reacted to a large extent with the contaminating plant material and most of the resulting hybridoma cell lines produced antibodies against plant constituents. The reason for this is that mice (the animals usually used) are very sensitive to sugars, like arabinose, that are found in plant material. In such cases considerable screening of cell lines is needed to find those producing the required antibodies.

panning for B-cell isolation

There are some methods available to improve the ratio of wanted/unwanted cell lines. The first one is the enrichment of immunoglobulin-producing B cells by 'panning'. For this, the target antigen is fixed to a solid phase, (eg on the bottom of a Petri-dish) and the B cell suspension floated over it, B cells exposing the right immunoglobulins on their surface will be bound and after washing away the unbound cells, the specific immunoglobulin-producing B cells can be harvested and fused with appropriate cancer cells. The method has already be used successfully to produce monoclonal antibodies against plant viruses.

In the second method, new-born mice are injected with the cross-reacting material (ie plant materials) to make them tolerant to this material. Later on the mice can be injected with the target antigen contaminated with cross-reacting material and then they will only produce antibodies against the target antigen. This method has been tried successfully for soluble antigens of plants. In preliminary experiments with antigens on the surface of bacteria the method has however so far failed.

10.4.2 Polymerase chain reaction

We remind you that if the base sequence of a DNA is known it is possible to make primers to both strands and to produce duplicates of the strands by extending the primers with a polymerase. Each time the procedure is repeated the amount of strands is doubled. If one uses the thermostable Taq polymerase the whole procedure can be automated and one duplication cycle can be done in about 5-7 minutes. The procedure will result in multiple copies of the DNA strands between the two primers. In theory 2 copies will result in 2^{21} copies after 20 repeats of the procedure. In practice it is usually loss due to a variety of reasons. The amplified material can be analysed by standard methods used to detect nucleic acids, (eg electrophoresis and molecular hybridisation). The great advantage of the method is its sensitivity. This is also its biggest disadvantage: one has to work very carefully otherwise one can detect everything everywhere!

The automated polymerase chain reaction method was first used in 1988. It is now used routinely in many fields (eg to increase the amount of papilloma virus for the detection of human cervical cancer).

In plant pathology the method has already been used to detect *Spiroplasma citri* (a mycoplasma) in extracts of *Catharanthus roseus*. In this case the gene for spirulin, a specific protein in the cell wall, was amplified. It was claimed that 10 fg of DNA could be detected. This would correspond to one *S. citri* cell. The method has also been tried for the detection of *Erwinia chrysanthemi*. In the latter case the very specific genes for pectate lyases were amplified.

detection of RNA using reverse transcriptase
In principle the method can also be used for the single-stranded RNA of plant viruses. The only problem is that the RNA has first to be transcribed to DNA by a reverse-transcriptase reaction. With purified RNA this can be done by adding the reverse transcriptase to the reaction mix for the polymerase chain reaction and a small adjustment in the programming of the computer that is controlling the reaction processing. It is not yet clear whether or not plant material will negatively influence the reverse-transcriptase reaction.

Interference by contaminating DNA is a problem. The polymerase chain reaction efficiency is negatively influenced by the presence of other DNAs in the reaction mixture, and therefore it will be difficult to use the method quantitatively.

10.4.3. Discussion

standardisation of reagents used in diagnosis
Biotechnology can be used to improve the efficiency of the production of monoclonal antibodies against plant pathogens. Due to the availability of monoclonal antibody producing cell lines that can be stored for a long time and used again to produce new batches of monoclonal antibodies that are identical to the ones produced years before, it will be possible to a large extent to standardise detection methods. In all parts of the world, now and in the future, tests can be made with the same ingredients and therefore with the same reliability.

The polymerase chain reaction will help to increase very small amounts of genetic material to easily detectable amounts and so will enable us to detect the smallest entities of pathogens (ie a single bacterial cell or a single virus particle). It may be used on a large-scale basis because the polymerase chain reaction itself can be fully automated. Because it is based on the detection of nucleic acids, it can also be used to test for the presence of viruses in transgenic plants.

To fully exploit the new biotechnological possibilities a lot of creative research is still needed. However, it will certainly lead to elegant new methods.

Summary and objectives

In this chapter we have examined the methods available for detecting the presence of pathogens in plant extracts. Particular emphasis has been placed on the modern molecular based approaches especially the use of antibodies and molecular hybridisation.

Now that you have completed this chapter you should be able to:

- list the advantages and disadvantages of conventional bioassay procedures for detecting pathogens;

- describe the advantages and disadvantages of molecular hybridisation and serological methods for detecting viroids, viruses, bacteria and fungi;

- give examples of how biotechnology may contribute to the development of diagnostic procedures used for the detection of plant pathogens in the future.

Responses to SAQs

Responses to Chapter 1 SAQs

1.1   There are several ways in which you could have arranged the boxes. We have chosen the one shown below. At the top, we have placed the box containing 'New plant strains' since the other boxes are all concerned with generating new plant strains. The box containing 'Gene probes' is rather special since gene probes themselves do not generate new strains, but are concerned with screening the products of the other boxes in order to identify the desired stains. It would be worthwhile copying this diagram and pinning it up where you are studying because it gives an overview of the remaining section in this text.

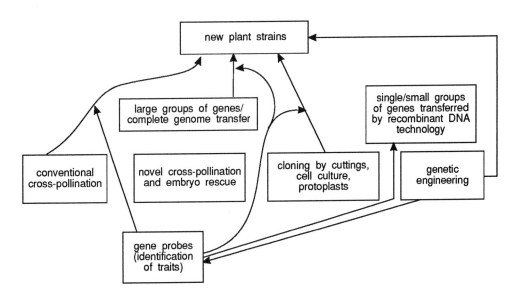

1.2   The correct order is

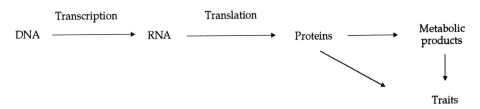

The information in the genetic material (DNA) is first converted into a nucleotide sequence in RNA by a process called transcription. The RNA nucleotide sequence is then converted (translated) into the amino acid sequence found in a protein. Many proteins act as catalysts (enzymes) which catalyse the synthesis of metabolic products.These metabolic products (or proteins themselves) bestow upon the plant its particular characteristics (traits).

	Conventional cross-pollination	Recombinant DNA technology	Cell fusion
a) This technique involves the transfer of a single or only a few genes		√	
b) This technique only applies to genetic recombination between closely related plants	√		√
c) This technique cannot be applied to developing traits which are the products of many genes		√	
d) This technique can be used to introduce genes from bacteria		√	
e) The outcome of this technique are highly unpredictable	√		√

1.3

a) Recombinant DNA technology involves the transfer of one or, at most, a few genes. The other techniques involve large blocks of genes.

b) You may have only ticked conventional cross-pollution. Although cell-fusion can be achieved between cells from quite unrelated plants, experience has shown that only the fusion of cells from quite closely related plants produce viable off-spring. Thus we are tempted to give cell-fusion half a tick!

c) Recombinant DNA technology, which involves only limited (single) gene transfer cannot be used to transfer traits which are multi-genic. The other techniques, which result in multigene transfer, are more plausible routes to changing characteristics.

d) It follows from the response we gave to b) that the only one of the techniques described which can be used for the transfer of genes from unrelated organisms is recombinant DNA technology.

e) Cynics might have put a tick against all of these. The most likely technique to give rise to a predictable result is the technique which involves the transfer of a single gene (ie recombinant DNA technology). It could be argued that we have so much experience with conventional cross-pollination that the result is largely predictable. The result of such an approach is however a large collection of different strains showing a variable combination of traits. Only after careful selection from this large variety of products can we isolate the plants with the features we desire.

Responses to Chapter 2 SAQs

2.1   There are many properties you could have cited for this SAQ. The main ones we expected you to list were:

   •   cultivated plants must flower at the same time as the related wild flowers if they are to cross-pollinate them. If the transgenic plants flower at a different time than their wild relatives, there is little prospect of the transgene(s) being transferred into the wild plants;

   •   if a transgenic plant is to cross-pollinate with wild flowers, the relevant (related) wild flowers must be in the vicinity of the transgenic plant;

   •   there are no natural plant viruses that are known to act as vectors between plant varieties.

Of course mankind can do much to reduce the chance of transgenic transfer. This is the point of SAQ 2.2.

2.2   Your list should have included all the devices that you could think of which prevent parts of the transgenic plants (especially the pollen) reaching their wild relatives. In other words, preventing out-breeding.

Thus, your list should have included:

   •   ensuring that wild relatives are not present in the immediate environments of the transgenic plants;

   •   reducing transfer of pollen into the environment by restricting (preventing) access by pollen transferring insects or by operating in containment facilities (eg in growth cabinets or greenhouses with constrained or filtered out flow);

   •   using good plant hygiene practices by, for example, destroying plants after completion of the field tests.

As part of the background studies before undertaking field trials, it is important therefore to evaluate which plants in the area are likely to be fertilised by the transgenic plant.

2.3   1) True.

   2) True, but the answer needs qualification. Viral resistance is often mediated by the production of viral coat proteins. Introduction of viral resistance into new crop plants will reduce the propagation of the virus. This in itself is unlikely to have a major impact on the environment. If however the virus also infects wild relatives, the reduction in the rate of production of viruses could have an indirect effect on the population of wild relatives (ie increase their chance of survival).

   3) True - the difficulty arises because the type of bactericidal proteins cited are not highly specific and because the longevity of these proteins in the environment is

unknown. The effects of these proteins on normal soil flora, or on the micro-flora of the guts of consumers of the plant (eg insects, animals, mankind) are uncertain. For transgenic plants making such products much work needs to be done to evaluate their effects on the environment and mankind.

4) False - but again the answer needs qualification. Chitinases are common enzymes. To act they must come into contract with their substrates. The most likely prolonged contact within the growing plant is in the hyphal walls of fungi invading the plant. Therefore the most likely consequence of transgenic plants producing chitinase is improved resistance to fungal invasion. The other major source of chitin in nature is in the exo-skeleton of insects. There are relatively few examples where prolonged contact between plant and the exo-skeleton of insects can be visualised. The exception is, however, found where the insects use plants (or part of plants) for physical shelter. A good example is provided by the gall-causing insects. The conclusion is therefore that introduction of genes into plants will be largely specific in restricting fungal spread but that it may have some effect on some insects groups.

5) Probably false. Whether or not a proteinase inhibitor will have effects other than reducing the spread of insects which feed on plants producing the inhibitor depends on:

- the specificity of the inhibitor;

- the stability of the inhibitor;

- the nature of the plant and the uses it is put to.

   In the text, we cited the example of plants which may be eaten by mankind in an uncooked state. The effects of the proteinase inhibitor in this case depends on its specificity. If it inhibits human as well as insect proteinases, then there will be important consequences. Likewise, if the proteinase inhibitor is stable to cooking or is recalcitrant to inactivation this is also likely to have a greater environmental impact.

6) False. The key word in the statement was 'invariably'. In the text, we described how herbicide resistance could be used to encourage the use of environmentally-friendly herbicides. Of course introducing resistance to environmentally damaging herbicides into a crop would encourage the use of these materials. We also described the situation in which herbicide resistance might be transferred to wild species. This would render an environmentally-friendly herbicide ineffective. We conclude therefore that although introduction of herbicide resistance can offer environmental advantages, we must be cautious and careful monitor the consequences of introducing such resistance into plants.

2.4   The main reason why it is essential for breeders to receive a fair amount of revenues for new cultivars is that without such revenues, breeders will not continue to produce new cultivars. In other words the potential benefits of the new technology in plant breeding will not be realised. These benefits include higher yielding crops, better food value crops, cheaper food and environmentally friendly agricultural practices. Without a good return on investment, plant breeders could not afford to continue to try to achieve these objectives.

2.5

1) False. Atrazines are long-lived and persist in the environment. Development of cultivated plants carrying this resistance will encourage the use of these polluting herbicides.

2) True. Marrows have been cultivated in these areas for a long time with no evidence of them becoming established as weeds. The plants do not survive winter conditions. Introduction of the kanamycin genes do not produce any enhancement of the prospect of the plants surviving in the environment. There is no evidence that kanamycin genes would be transferred to other organisms (especially bacteria).

3) False. Although many carrots are cooked before eating and this would probably destroy the inhibitor, some carrots are eaten raw. Thus, this broad spectrum inhibitor may well inhibit human gut proteinases.

4) True. The cost of the development of transgene plants tends to restrict the range of cultivars that are worked upon. The dangers of genetic erosion are recognised and much effort must be put into developing gene banks to ensure that genetic variety is not lost.

5) False. The most likely people to benefit are those living in temperate climates. Those who grew and marketed the product from tropical regions will probably lose most (or all) of their markets for the crop. This will, of course, face them with considerable economic hardship.

2.6

We will use the headings of the risk analysis described in 2.5.1 to provide an answer to this SAQ.

1) Description of the plants used.

Grass is a perennial crop. It persists and can spread.

2) Potential harmfulness of the transgenic plant.

Although grass is harmless to many organisms the introduction of proteinase inhibitor genes may affect those animals and insects which graze on it. This includes both domestic (cows, sheep, horses) and wild (rabbits) animals.

3) Growth and survival

The introduction of the inhibitor genes is likely to dramatically increase the survival of the grass strain.

4) Genetic stability

There are also substantial amounts of cross-breeding between grass strains.

On the basis of these few features alone, do you think that the proposed product offers a high or low environmental risk? Of course the answer is high because the weediness of the plant is increased and it offers hazards to a wide range of fauna. It would not be sensible to proceed with the project.

Responses to Chapter 3 SAQs

3.1

a) diploid (2 sets); homozygous (2 sets of identical chromosomes).

b) triploid (3 sets); homozygous (chromosome sets are identical).

c) diploid (2 sets); heterozygous (we shaded one set of chromosomes to indicate that they are different).

d) aneuploid - This was a little difficult, there are two sets of all of the chromosomes plus two extra copies of chromosome 4.

3.2

1) This applies to the plants. Since they are all derived from one plant, they will contain the same sets of chromosomes and genes. They can therefore be described as genetically homogenous.

2) This does not apply. The new plant was produced by the combination of chromosomes from two genetically distinctive parents. In other words it would be heterozygous not homozygous.

3) This applies to the population of plants. Since they are produced vegetatively from a single plant, they will all contain the same genes (ie the plants would be genetically identical). We call such a population a clone.

4) Does not apply. Since the plants are incapable of interbreeding, they can hardly be described as an interbreeding group. Left to themselves, the plants would become extinct in a single generation unless they could propagate themselves vegetatively.

5) Providing the properties of the population could be described in sufficient detail and thus allowed the population to be distinguished from other populations, we could describe the population as a cultivar.

3.3

2), 3) and 5) are most likely to show quantitatively inherited variation. In many cases resistance to viral infection 1) may be attributed to one (or a small number of linked) gene(s). Likewise, resistance to a herbicide 4) can be attributed to a single gene. As far as these phenotypic characters are concerned, the progeny either have or have not got the relevant gene. Thus they will either be resistant or sensitive to the virus or herbicide. There would be no intermediate stages (ie partial resistance). These are examples of qualitative inherited variation.

Cold tolerance 2), plant size and shape 3) and seed number and size 5) are each usually controlled by a variety of genes (ie they are polygenic). These characteristics usually show a continuum between two extremes and are thus examples of quantitatively inherited variation.

3.4

1) H = 0. We would expect the environmental variation to be minimal (ie Ve - 0) since the cabinet provides a consistent environment for each plant. We would also expect genetic variance (Vg) to be 0, because all of the plants are genetically identical (ie a clone).

Thus since $H = \dfrac{Vg}{Vg+Ve}$ then $H = 0$

2) $Vp = 0$. Since $Vp = Vg + Ve$ and we have already established that Vg and Ve are both $= 0$.

3) Vp would increase because the plants would each be in different environments (eg against a wall or next to a path). This would lead to an increase in Ve. Therefore Vp would increase.

4) $H = 0$. Although there would be an increase in Vp, the variation arising from genetic variance (Vg) would still be equal to 0 (we are still cultivating a clone).

Since $H = \dfrac{Vg}{Vp}$ then $H = 0$

3.5

1) N - gene Nm is recessive, N is dominant, therefore the product will be N.

2) Nm - although Nm is recessive, no N gene is present.

3) Nm - since Nm is dominant, in the hybrid NNm, Nm will be produced.

4) Nm - this is homogenous for Nm, therefore Nm must be the product.

3.6

To answer this you will need to have remembered what we said about hybrid varieties or cultivars towards the end of section 3.4.2. Hybrid cultivars are important commercially. If a hybrid is allowed to interbreed, gene/chromosome combinations that are not so desirable will be generated. Male insterility will, of course, prevent this interbreeding.

3.7

This question requires quite an extensive answer. There are many possible ways in which biotechnology may be applied. First the biotechnological *in vitro* multiplication of plants and plant parts and the production of clones enables us to produce large numbers of homozygous parents. Secondly we may be able to use genetic probes (eg cDNA) to identify suitable parent strains (ie those carrying a particular gene). Thirdly, genetic manipulation may enable us to introduce new, desired genes, into one of the parents. Fourthly, protoplast fusion (somatic hybridisation) may enable us to introduce new combination of blocks of genes. Fifthly, we may use somatic fusion techniques to introduce cytoplasmically inherited male sterility genes into the cells thus preventing hybrids from inbreeding. Finally we could use the *in vitro* multiplication technique to produce clones of the hybrid for sale.

You may or may not have spotted all of these. You may also have included some additional ones. The point we are making is that biotechnology can make major contributions to plant breeding in a wide range of areas. By the time you have completed this text, you will be armed with much more detail of this contribution.

Responses to Chapter 4 SAQs

4.1   You should have selected all six, but they each need some expansion.

1) Encourages the use of tissue culture techniques because the higher the ploidy state the greater the difficulty to achieve genetic segregation and to produce a uniform product (ie the plants remain genetically heterogenous and therefore phenotypically heterogenous).

2) Tissue culture techniques enable the producer to ensure that the propagules are free of disease. If the plant has no known disease then cultivation in expensive controlled, aseptic environments is largely a waste of resources.

3) This is really the phenotypic consequence of the condition described in 1). Growers usually prefer to produce predictably homogenous crops since this enables implementation of cost-effective cultivation and harvesting techniques. Cloning by tissue culture techniques obviously achieves this.

4) This is perhaps self evident, the greater the advantage placed on vegetative propagules, the greater will be the premium paid for their production.

5) This should be obvious! A high price for a product always encourages its production.

6) Is a little more complex, but usually restrictions on the transfer of plants are usually based on reducing the introduction of non-indigenous diseases into an area. Guaranteed disease - free plants would, of course, enable relaxation of such regulations.

The main argument against using tissue culture techniques compared to seeds to generate propagules is that the former usually demands many more resources (eg equipment, skilled personnel) which tends to make the products of tissue culture rather more expensive.

4.2   1) Tubers are effectively dormant, storage organs. Plantlets transpire and therefore need a constant supply of water. They also tend to be tender and likely to be damaged by physical and environmental stress.

2) The main differences between *in vivo* and *in vitro* cultivation may not be obvious from the diagrams. *In vitro* cultivation takes place in an environment free from pathogens often using chemically defined media. *In vivo* cultivation takes place in natural environments (ie pathogens and other micro-organisms may be present) often using chemically undefined growth media (eg soil). We can however spot some specific differences. For example *in vivo* cultivation of plants using flowers means that the usual propagule that is produced is a true seed produced by self- or cross-pollination. In *in vitro* cultivation using the flower, usually only part of the flower (eg anther, microspore, ovule, embryo) is used as the propagule. The *in vitro* methods have the potential for large (infinite) multiplication. This is symbolised on our diagram in the text by our recycling arrows on each of the *in vitro* procedures.

4.3   Frost, heavy winds, drought (especially when combined with strong wind, high temperature or high light intensity). Plantlets are sensitive to environmental stress.

4.4   The costs of production are higher, because of:

- the need for changes in the medium;

- the slow growth of tubers *in vitro*;

- the low number of tubers per plantlet, so that little multiplication takes place during the *in vitro* tuberisation.

4.5   Meristems are preferred for cryopreservation, because they are: genetically stable; are very small; usually do not contain viruses or other diseases; are fairly easy to treat with cryoprotectants; have a high survival rate and are easy to regenerate.

4.6   The seeds are large because they carry a large stored food reserve to nourish the germinating seedling. This is usually absent from somatic embryos especially those species in which the seed depends upon a large endosperm reserve. Provision of an equivalent energy source in the gel or capsule surrounding the embryo would be an ideal substrate for micro-organisms when the embryos were planted. Infection of this sort might lead to high rates of mortality amongst the embryos. Note however that some embryos do contain seed reserves especially in those in which the cotyledons act as a storage organ.

4.7   1) aneuploidal - this is the term used to describe the loss or gain of individual chromosomes.

2) euploidal - the plants appear to have the same number of chromosomes, but they obviously carry different genetic information regarding flower colour.

3) chimeric - since flower colour is governed by genes, and since different parts of this plant were producing differently coloured flowers, it appears that the different parts of the plants are genetically different.

4.8   1) c) is the correct answer. Meristems a) are usually small and difficult to isolate. Unopened flower buds b) and thick walled seeds d) are not easy to treat with enzymes. Leaf mesophyll cells are usually readily obtained in large quantities and are open in structure thereby allowing ready penetration by cell wall degrading enzymes.

2) a) and c) are correct. Meristems are capable of rapid growth and subsequent differentiation and are therefore suitable for micropropagation. The ability to grow, low levels of contamination by disease (viruses) and small size make them suitable for cryopreservation.

3) d) is correct.

4.9   Your flow diagram should look something like the one below

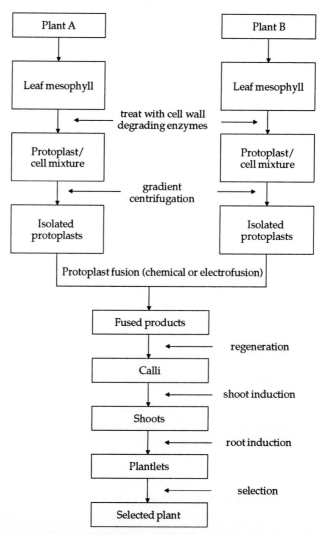

4.10   Somatic embryogenesis and normal (zygotic) embryogenesis. Direct organogenesis is the best method for rapid multiplication, because it does not affect the genetic stability. It does not give the highest rate of multiplication. Highest rates are obtained with somatic embryogenesis in those species where it is possible.

4.11   In principle the number of shoots per tuber can be predicted and controlled. This control is important for the yield and size distribution of tubers. Also important is that earlier onset of growth and therefore a higher yield occurs with sprouted tubers.

4.12   A faster early development results in an earlier closure of the canopy. A closed canopy intercepts almost all the incoming radiation, thus limiting the growth of the germinating weeds. The closed canopy also reduces the heating of the soil by radiation. The soil temperature is very important for the occurrence of physiological disorders. A high temperature (eg because of a lot of incoming radiation) increases the occurrence of such disorders.

4.13   All cultivation techniques that increase the rate of early growth. Early growth can be enhanced by the use of pre-cultured transplants instead of tubers, use of mulches that increase the soil temperature early in the season, for example transparent plastics (see Figure 4.10) and sufficient water supply, especially during early growth.

Responses to Chapter 5 SAQs

5.1   1) Variants. Although there are obvious differences between the two types of plants, no evidence is provided that the differences are a product of changes in the genes. The differences could be due to epigenetic changes. It would be difficult to prove that the difference between the plants was genetic because the defective flowered variants would not produce gametes which could be used for the sexual generation of zygotes (ie we could not prove if the variants were transmissible through meiosis).

2) The closer examination of the cells from the two variants reveal chromosomal differences. Therefore it appears that the variants were probably based on genetic differences. The defective - flower variants seem to be aneuploids.

5.2   1) B. Plants which carry recessive genes would expect to produce about ¼ of their offspring as homozygotes for the recessive gene when they are self-fertilised. Genetically this is written as:

Ff x Ff

1FF + 2Ff + 1ff

This rate of recessive homozygotes (ie yellow flowers) is given for plant B.

2) It appears that the gene for flower colour is unstable. One (plant B) of the five (A-E) original plants was found to carry f allele despite the fact that all were derived from FF genotype plant. This suggests that F often mutates to its recessive form.

3) We have concluded that B is a heterozygote (Ff). E appears to be a homozygote (FF). Therefore if they are crossed, then:

FF x Ff

FF + Ff

50% of the progeny would be homozygous (FF) and 50% heterozygous (Ff). All would contain the dominant gene. Therefore all the flowers would be red. However, we argued in response to question 2), that F is unstable and that it often changes to f. If this is the case, then a few of the offspring from the FF x Ff cross, might produce yellow flowers (ie have the genotype ff).

5.3   Insertion of a transposable genetic element into A will inactivate this gene. We can draw such a cell.

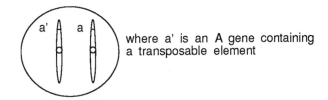

where a' is an A gene containing
a transposable element

This cell would of course not show the phenotypic characteristics of expression of the A allele. Such changes would also be present in the progeny of these cells until the element was excised from the allele (reversion).

5.4   1) True. The mobility of transposable genetic elements and their insertion into functional genes, leads to changes in the genetic information. We call such changes mutations.

2) False. The opposite is true.

3) True. Before growing as calli the differentiated cells have first to become de-differentiated. This process appears to favour the production of mutant cells.

4) False. Although mutations may lead to stable changes in gene expression, such changes can be achieved in other ways. For example in a partially differentiated cell, the changes in gene expression may persist in its progeny. Such changes we call epigenetic.

5) True. Observation of the chromosome number using a light microscope readily reveals chromosome number. This number can be used to identify changes in ploidy and in aneuploidy. Small mutations within a chromosome cannot be visualised by microscopy. They are only easily identified if they bring about some detectable phenotypic changes.

5.5   This is quite a tricky set of questions. Let us deal with each in turn.

1) For an over-producer of F, it must mean that we have a mutation in the gene converting A $\rightarrow$ B so that F no longer inhibits its own production. In a normal cell, if we add J and N to high levels, H will accumulate. This will inhibit the conversion of A $\rightarrow$ B. This would therefore prevent F from being made and the cell would not grow. If the gene converting A $\rightarrow$ B was mutated so that it was no longer sensitive to H. A would be converted to B and this to F. The cell would therefore grow in the presence of excess J and N. If the sensitivity of the A $\rightarrow$ B converting enzyme to H was changed, it could be that it was also insensitive to F. In other words there is a high probability that cells tolerant to high levels of J and N would be over-producers of F.

2) The answer is no. If we used F as a marker for a change in the gene coding for the enzymes converting A $\rightarrow$ B, then J and N would still inhibit their own production (ie J would inhibit H $\rightarrow$ I and N would inhibit H $\rightarrow$ M). Therefore an F over producer need not be an J or N over producer.

3) The answer is probably that F production would be unaffected or even decreased. Over production of N would suggest that the H → M step was no longer sensitive to feedback inhibition. Thus there would be greater flux through the pathway A → B → C → G → H → M → N. This may mean that less C was available to be converted to F.

These questions will show you how a detailed knowledge of metabolic pathways and their regulations can be helpful in designing selection strategies.

Responses to Chapter 6 SAQs

6.1   The ploidy level of the sexual hybrid will also be diploid. In sexual hybridisation of diploid species both parents contribute one set of chromosomes by producing haploid gametes.

6.2   The somatic hybrid plant should be male and/or female and fertile. This will enable the desired genes to be passed on to the progeny.

6.3   Membrane breakdown may result in membrane fusion of two adjacent protoplasts or in cell lysis. Breakdown of the membrane, outside of the membrane contact area causes cell lysis. If a very high DC voltage is applied then the chances of cell lysis increase. Longer pulses have the same effect.

6.4   There are many points you could make in answer to this question. We cite the following:

PEG fusion	Advantages	no electrical equipment needed
	Disadvantages	agglutination of protoplasts; cytotoxic effects; low fusion frequency
Electrofusion	Advantages	rapid; reproducible; high fusion frequency; can be followed under the microscope
	Disadvantages	special apparatus is required

6.5   The fusion products are fragile directly after fusion. After 1 to 2 days cell walls are being formed and fusion products will better withstand mechanical manipulation ie collection with a micro-pipette.

6.6   Auxotrophic cells need a special compound to be added to the culture medium to ensure cell proliferation. After fusion with a non-mutant cell, metabolic complementation occurs. Thus, hybrids can be selected on a medium lacking the special compound but to which the drug is added. The drug will inhibit growth of the non-mutant cells and the mutant cells will be able to divide on the medium without the special compound added.

6.7   1) Both X and Y are hybrids of cell lines 1 and 2 since both contain isoenzymes present in both 'parent' lines.

   2) Nothing can be concluded with any great certainty. Nevertheless X appears to produce the complete set of isoenzymes from both parent lines. It would appear therefore that X contains 2 complete sets of chromosomes. Thus if the parent cell lines were haploid, X would appear to be a diploid. If on the other hand cell lines 1 and 2

are diploids, then X is probably a tetraploid. We cannot be certain of this because we do not know if all of the chromosomes of the parent lines are represented by the isoenzymes represented in the drawing.

Line Y is probably an aneuploid. Although it produces many of the isoenzymes of each parent cell line, it does not produce them all. For example it appears to only produce cell line 2 A isoenzymes. This could be explained by the loss of a chromosome(s) from cell line 1 following fusion.

3) These observations could be conclusively proven by microscopic observation and counting of the number of chromosomes present.

6.8 The donor protoplasts cannot survive because their nuclei are inactivated by the radiation treatment. The recipient protoplasts are metabolically inhibited (poisoned) by the iodoacetate and therefore are unable to support growth.

6.9 With the donor-recipient system using irradiation of the donor, the product of fusion contains a mixture of cell types including those with complete and incomplete donor chromosomes and donor chromosome fragments integrated into the recipient chromosomes. Also other types of radiation damage may be present. With the microprotoplast system these phenomena do not occur. It is expected that with this system a few complete chromosomes can be transferred.

Responses to Chapter 7 SAQs

7.1

1) ffss - Since white flowers and wrinkled seed are both recessive; to express these two phenotypic characters, the plant must be homozygous for these two genes.

2) There are several possibilities. The following genotypes will produce red flowered, round seed products: FFSS, FfSS, FfSs, FFSs.

3) Yes, the genes are linked. The homozygous red flowered round seed plant would have the genotype FFSS. The other parent would be ffss (see 1 above). Therefore these would produce the gametes: FS and fs respectively.

Therefore the genotype of the F_1 generation would be FfSs. If the genes were unlinked then the following gametes would be produced: FS, Fs, fS and fs.

Crossing these would give the following genotypes in the progeny:

	FS	Fs	fS	fs
FS	FFSS	FFSs	FfSS	FfSs
Fs	FFSs	FFss	FfSs	Ffss
fS	FfSS	FfSs	ffSS	ffSs
fs	FfSs	Ffss	ffSs	ffss

Phenotypically:

- FFSS, FFSs, FfSs and FfSS would be red flowered, round seed;

- ffSS and ffSs would be white flowered, round seed;

- FFss and Ffss would be red flowered wrinkled seed;

- ffss would be white flowered, wrinkled seed;

- The ratio of these would be 9:3:3:1.

If the genes were linked then the gametes of the F_1 generation would be FS or fs. Therefore the F_2 generation would have the genotypes:

	FS	fs
FS	FFSS	FfSs
fs	FfSs	ffss

Therefore the progeny would either be white flowered and wrinkled (ffss) or red flowered and round (FFSS; FfSs). The ratio of these two phenotypes would be 3:1. Thus the genes are tightly linked.

4a) The small number of white flowered and round seed and red flowered, wrinkled seed in the progeny suggest a small amount of crossing over occurred. Thus instead of just producing FS and fs gametes a small number of Fs and fS gametes were produced. The combination of these gametes would produce a few FFss, Ffss, ffSS and ffSs genotype progeny. These show up as white flowered, round seed and as red flowered wrinkled seed progeny. The low numbers of these suggest that crossing over was a rare event and that the genes were probably closely linked.

4b) The arguments for this question are similar to those given for a) but in this case the greater proportion of recombinants suggest that crossing over occurred more regularly. Thus the flower and seed genes, although linked are probably not so closely linked as in the example given in 3).

7.2 To answer this question you will need to think logically. The parents genomes were FFss and ffSS. Therefore the gametes produced would be Fs and fS respectively. Therefore the F_1 generation would have the genotype FfSs (ie they would all be red flowered, round seed). These could produce gametes with genotypes of Fs or fS and a few fs and FS as a result of crossing over.

These were mated with a plant of genotype ffss (which would produce fs gametes). Thus the possible genotypes of this progeny would be:

	FS*	fs*	Fs	fS
fs	FfSs*	ffss*	Ffss	ffSs

Those with asterisks are products of crossed over chromosomes. These would have the phenotypes of white flowered, wrinkled seed (ffss) or red flowered, round seed (FfSs).

$$\text{The crossing over frequency} = \frac{\text{total number of recombinants}}{\text{total progeny}}$$

$$\text{From the question} = \frac{24+22}{340+336+24+22}$$

$$= \frac{46}{722} \text{ ie approximately 0.064 or 6.4\%}$$

7.3 The lowest frequency will be those derived from double cross-overs. Thus AaBbcc and aabbCc represent double cross-overs (ie from genes ABc and abC). This suggests that the order of genes must be ACB.

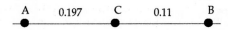

Let us check this. If the order is as described we could produce the desired gametes by a double cross ie:

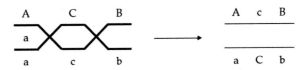

30 (= 14 + 16) double cross-over recombinants were produced. Thus the double cross-over frequency = 30/2000 = 0.015 (1.5%). NB total number of progeny = 690 + 726 + 170 + 194 + 14 + 16 + 86 + 104 = 2000.

The number of cross-overs between A and C is 170 + 194 + 14 + 16 = 394 (ie cross-over frequency = 394/2000 = 0.197 = 19.7% ≡ 0.197.

The number of cross-overs between B and C is 86 + 104 + 14 + 16 = 220 (ie cross-over frequency = 0.11 = 11.0%).

ie we can map these genes as:

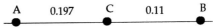

7.4   The autoradiograms of plants 1, 2 and 4 suggest that in these plants genes A and B are coded for on the same restriction enzyme fragment.

In plant 3 genes A and B appear to be on different fragments. In other words there is at least one restriction enzyme site between them. The probable sequence of events is therefore that the situation occurring in plant 3 may have arisen by mutation producing a new restriction enzyme site between the two genes. Thus:

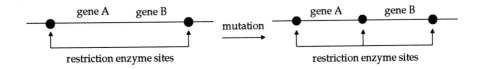

The observed result could however have occurred by some other mechanism eg inversion or insertion.

To determine whether or not inversion or gene migration had occurred would require further analysis using other restriction enzymes.

7.5   1) Sexual fusion of plants 1 and 2 led to production of progeny which appear only to contain gene A derived from the female. It appears therefore that gene A is maternally inherited. This is quite common for genes carried on extra-chromosomal (eg plastid) DNA. This provides us with several options for testing our conclusion.

2) If gene A is inherited maternally, we could expect that if we took pollen from plant 1 and used this to fertilise the ovaries of plant 2, the resultant progeny would produce autoradiograms similar to plant 2.

An alternative route to testing our hypothesis would be to extract extra-chromosomal and nuclear DNA separately. After restriction enzyme hydrolysis we would expect probe A to only hybridise with the extra-chromosomal DNA.

7.6   We know that the amino acid sequence is:

NH₂ met thr glu glu met trp ...... met trp thr val trp trp COOH

These can be coded for by the following nucleotide sequences

```
5'  ATG ACA GAG GAG ATG TGG          ..........     ATG TGG ACA GTC TGG TGG   3'

         T   A   A                                       T     G
                                                               A
         C                                               C     T

         set 1                                             set 2
```

Thus we should make two sets of primers. One set should contain a mixture of 12 different oligonucleotides (ie 3 x 2 x 2) to identify the start of the gene corresponding to the N end. The other should contain a mixture of 12 different oligonucleotides (ie 3 x 4) to identify the C terminal end of the gene.

How would these sets of oligonucleotides hybridise to the DNA? If you think about this carefully you will realise that they will both hybridise to the same strand.

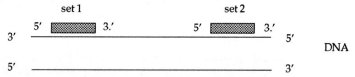

We really need to produce the complimentary strands to one of these sets. But which one?

If we use set 2, instead of producing: 5' ATG TGG ACA GTC TGG TGG 3' we should produce:

3' TAC ACC TGT CAG ACC ACC 5'

Using these and the original set 1 oligonucleotide, these would hybridise to the DNA thus:

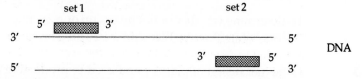

These two sets of oligonucleotides would act as primers for the synthesis of both strands. Remember that DNA is synthesised from 5' to 3'.

7.7   The trick here is to make a direct comparison of the autoradiograms. This could be done by making a tracing of the autoradiograms produced using root mRNA. This shows all the plaques which contain DNA which is expressed in roots. If we lay this trace on the other two autoradiograms, we can identify those plaques which contain DNA which is also expressed in leaves and flowers. These are indicated below as:

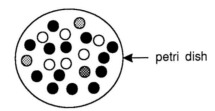

●   flowers, leaves and roots
◉   flowers and roots
◎   leaves and roots
○   root specific

← petri dish

7.8

1) b), c), e) and f). The clue is in the loops of genomic DNA. This indicates that after the nucleotide sequences in the DNA are transcribed, some are excised during mRNA processing. They are removed from the mRNA and are therefore missing from the cDNA.

2) At least 6 - all 6 illustrations appear to be of related genes although all have different introns. They all hybridise with the same cDNA. There could, of course, be even more but they have sufficiently different nucleotide sequences such that the cDNA no longer hybridises with them.

3) d) Note that the cDNA hybridises with two separate genomic DNA fragments. In this case we know that the genomic DNA contains at least one restriction enzyme site within the gene.

7.9

This is quite a difficult question. Let us see if we can be systematic.

Plant A and B were both homozygous. From the results of the crossing it appears that red colour is dominant. Thus we can write the cross as:

Plant A	x	Plant B
rr		RR
F₁ progeny	Rr	Red flowered

The one white flowered plant in the F₁ progeny probably has a mutation in the R gene. A common cause of mutation is the introduction of a transposon in the relevant structural gene. Using the cloned radioactive transposon, it appears that each F₁ plant carries transposons on 3 restriction enzyme fragments. But there is one of these in a different position in the white plant. This could be R gene with the transposon inserted. If we eluted and analysed this fragment we would expect it to have the structure:

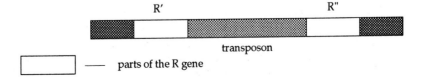

R'   R"

transposon

☐ — parts of the R gene

This could be cloned in a vector and prepared in a radioactive form. If we now took the DNA from a red flowered plant, treated this with a restriction enzyme, we could use the cloned R'-transposon-R" DNA as a probe for the R gene. Once we had located this on the electrophoretogram of DNA from red flowered plants, we could elute the DNA and clone it. This would contain the R gene without an imposed transposon.

Responses to Chapter 8 SAQs

8.1   The main disadvantages are:

- barriers to genetic transfer between species;

- reassortment of all traits (good or bad) in the progeny;

- time consuming back-crossing programme.

8.2   Advantages are:

- add isolated trait to existing variety;

- fast procedure.

Current limitations are:

- not all plants can be transformed;

- tissue culture methods inadequate for many crop species;

- number of genes limited;

- expression of introduced genes not always understood.

8.3   Introduction of T-DNA into the plant genome. In the case of *Agrobacterium tumefaciens* the T-DNA contains genes encoding enzymes involved in the biosynthesis of growth hormones auxin and cytokinin. These hormones stimulate cell growth causing tumour formation. Although it was not described in the main text, in the case of *Agrobacterium rhizogenes* the T-DNA encodes proteins modifying the sensitivity of the cells for plant hormones possibly by modifying the hormone receptors or the hormone transport.

8.4   For the production of opines which can be catabolised by the bacterium. Thus the bacterium causes the plant cells to make a product they can use as a substrate to grow.

8.5   The plant produces inducers like acetosyringone which modified the virA protein. Subsequently the virA protein phosphorylates the virG protein which then can induce other vir genes involved in the transport mechanism of the T-DNA. Before transfer can take place gene products of bacterial chromosomal genes chvA and chvB are necessary for attachment of bacteria to plant cells. The bacterial virB proteins may be involved in formation of a pilus or pore for transport of the T-DNA which is encapsulated in virE protein. The T-DNA is cut out of the T$_i$ plasmid by virD proteins. virE proteins appear to coat the T-DNA strands which are transported into the plant cell.

8.6   T-strands are formed after nicking of the bottom strand of the T-DNA right border sequence by virD proteins. This nick induces the repair mechanism of the host cell which means new DNA synthesis by DNA polymerase in the 5' to 3' direction. Thus the T-strand is replaced by the newly synthesised DNA from right to left border sequenc⌐

This is achieved by endogenous DNA polymerase. Release of the T-strand is accomplished by a second nick at the 3′ side by an unknown mechanism. Most likely virD proteins are involved.

8.7 No, a useful promoter has to be active in de-differentiating cells otherwise there is no mechanism for selecting transformed cells from the untransformed/transformed cell mixture. Thus a promoter insensitive to tissue and differentiation process (also called constitutive) is needed.

8.8 Yes. If a flower specific promoter is used to drive a reporter gene (eg GUS), we would expect that this gene would only be expressed in flowers. Such a promoter would, of course, not lead to gene expression in other tissues.

8.9 After injection the DNA can move to the nucleus and be integrated in the genome. However if injected into the vacuole of the cell it will be degraded. Injection into mitochondrion or chloroplast presumably could lead to incorporation if these organelles can survive injection damage. However successful organelle transformation has not been reported.

8.10 No, rapeseed can be easily transformed by *Agrobacterium tumefaciens*.

8.11 Most biochemical pathways show several reaction steps, which are each catalysed by a separate enzyme. Each enzyme is coded for by one or more genes, dependent on the number of different subunits involved. Some of the enzymes needed for a given pathway might already be present in the acceptor plant species. In general however, several genes are needed to confer an entire new pathway.

8.12 The expression of a single viral gene (eg the coat protein) will not lead to symptoms, and spread to more sensitive species can not occur. The coat protein gene in the plant may undergo mutations that render it less effective in its protection. But this will occur at a much slower rate than for the rapidly evolving viral genomes (think about the rate of DNA replication in viruses compared to plant cells). A disadvantage that can not be overcome is the limited protection range resulting from the presence of the coat protein. It would only protect the transgenic plant carrying the gene.

8.13 Yes. The fact that the intracellular receptors must be virus specific can be deduced from the observation that only the replication of closely related viruses can be inhibited by the presence of viral coat proteins.

8.14 1) Unlikely since the antisense code will contain a randomly located start codon that in most cases will lead to inefficient translation initiation due to the absence of proper translation regulatory signals. Stop codons may be rapidly encountered, leading to the production of short peptides.

  2) For antisense activity transcription is required to generate many molecules of antisense mRNA, so a functional promoter region is essential.

8.15 When several simultaneous insect pest infestations take place. These cannot be controlled by the same *B. thuringiensis* isolate.

8.16
- the stomach pH of mammalian species is acidic which leads to rapid degradation of the toxic fragments;

- larval epithelial cell wall receptors are highly specific even within the insect kingdom and sufficiently related receptor proteins probably do not occur in mammals;

- the proteases of mammals probably have different specificities to the insect proteases, therefore processing the protein in a different manner.

8.17 Since repetitive applications of *B. thuringiensis* in its use as a bio-insecticide are needed for effective control it can be concluded that propagation and survival of this bacteria in the field must occur at low frequencies. This is indeed what is observed after application of the bio-insecticide.

8.18 NPTII fusions can only be used when the protein product of the gene fusions retains both NPTII activity and the activity of the desired trait. This will be determined by the interference of the two protein domains together constituting the fusion (chimeric) protein. It can be easily envisaged for instance that the proper three dimensional folding of each domain in the engineered fusion protein could be obstructed.

8.19 The transgenic plants produced up until now only contain a single crystal protein gene and thus express only one protein, whereas most isolates contain several expressed crystal protein genes coding for crystal proteins exhibiting different, often partially overlapping ranges of insecticidal activity. The range of insects affected by an isolate is the sum of these individual crystal proteins.

8.20 The fruit must be eaten by birds and insects or disintegrate under fungal or bacterial action in order to stimulate spread of the seeds. Potent proteinase inhibitors may prevent consumption by insects and birds either directly by interfering with the metabolism of the individual or, in long term evolution, by selection for those individuals tending not to eat the fruit.

8.21 The newly introduced gene codes for a mutated enzyme analogous to the resident copy, it does not code for a function that blocks the expression of the resident enzyme. The resident gene will thus be normally expressed, although it codes for an enzyme the function of which is completely inhibited.

8.22 Bacteria can be easily grown and handled in large quantities. They show very short doubling times, optimally even less than one hour, which makes them excellent organisms in which to search for mutants. This is only possible if the protein products encoded by the genes of interest can be made to function both in prokaryotes and in eukaryotes.

8.23 Tolerance describes a phenomenon in which a process is still functioning despite the presence of an inhibitor simply by the high, advantageous ratio between target and inhibitor. The inhibitor is functional but at too low a level. In the case of resistance the action of the inhibitor is somehow eliminated.

8.24   Glutamine synthetase is also an essential enzyme in bacterial metabolism and like counterparts in other organisms is sensitive to the action of phosphinothricin which very much resembles the normal substrate glutamic acid. The bacterium simply needs the detoxifying enzyme PAT to survive.

8.25   It suffices to select for phosphinothricin resistant plants, so kanamycin resistance selection is not necessary.

8.26   Only one copy of the gene coding for PAT is needed to cause resistance, so in a crossing only one parent has to provide the gene.

8.27   Plant seeds need to be highly protected since they play an essential role in the sexual propagation of the plant and they sometimes need to be capable of survival over a long period of time. As for proteinase inhibitors, the abundant presence of the thionins may provide such a protection.

8.28   The main reasons we believe are that localised necrosis may provide a physical barrier to the infecting pathogen, and it may deplete pools of compounds needed for growth and spread of the pathogen.

8.29   Many pathogens spread through a plant through extracellular spaces. The secretion of PR proteins into these spaces may prevent the spread of the pathogen.

8.30   In the case of the red petunias, a new gene was introduced into the plant. In the other example, a resident genes expression was blocked leading to white flowers.

8.31   A multicellular origin would lead to the appearance of both homogeneous tubers in those cases in which all original cells showed comparable staining patters, and to heterogeneous tubers in the other cases. Since only homogeneous tubers were observed the tubers must have originated from a single cell.

Responses to Chapter 9 SAQs

9.1   1) Biological control agents are natural agents and many are specific for individual insects. They leave no residues and are environmentally safe.

2) Engineering to reduce UV sensitivity has the intrinsic problem that the viral DNA itself is UV sensitive (it forms thymidine dimers). The solution to this problem is most likely to be solved through spray formulations with UV protectants.

9.2   1) We can cite many reasons. Amongst the more likely are:

- inappropriate dissolution of polyhedra in the midgut;

- absence of appropriate receptors for virus attachment on midgut cell;

- failure of viruses to uncoat in the cell;

- no or aberrant transcription (promoters are not recognised);

- budding into the haemolymph impaired.

2) Occlusion bodies provide protection against environmental decay. It promotes the survival of the virus outside the host, which is a compensation for the lack of active transmission between individual insects. The possession of an occlusion body gives a strong selective advantage to survive. We can imagine therefore an initial, rapid evolution to produce structures which efficiently protect the virus. Then, when such structures have evolved, there is an evolutionary resistance to further change.

9.3   There are several options. The one that we favour is to extract the mRNA from an infected cell and produce a cDNA copy. This could be cloned and used as a probe.

If we now extract baculovirus DNA and hybridise this with the cDNA, the kinetics of hybridisation and the amount of hybridisation that takes place would give a measure of the number of copies of the gene in the viral genome.

Evidence that translation efficiency is not particularly high would come from the amounts of mRNA present in cells compared to the amounts of the protein product. These could be compared with other genes and gene products.

9.4   Ecdysteroids are compounds present in the haemolymph of insects, which initiate the moult in insect larvae. After the moult the ecdysteroids are inactivated by conjugation with glucose, mediated by a UDP glucosyl transferase. The instar continues to grow and produces large amounts of OBs. UDP glucosyl transferase is part of the developmental regulation of insect moulting and morphogenesis. The acquisition by baculoviruses of a UDP glucosyl transferase gives baculoviruses the possibility to block moulting at an early stage of the infection to the benefit of its own reproduction.

9.5   There are several possible reasons. We identify the following two as perhaps most important (remember we specifically asked for molecular reasons). To invade and infect the host, the virus must first attach to specific receptors on the surface of cells lining the gut. The absence of these receptors will mean the virus will be non-infective. Not all insects carry such receptors.

Secondly, to replicate and cause the symptoms of disease, the promoters of the viral genes need to be recognised by the host's RNA synthesis machinery.

9.6

1) Both baculovirus IE and host promoters are expressed under the control of host (transcription) factors, and are therefore equally effective in driving the expression of foreign genes. The level of expression of these promoters is much lower than of polyhedrin or p10, therefore less insecticidal activity should be expected.

2) Trypsin inhibitor would inhibit proteases from digesting the food.

3) *Bt* toxins are present in the digestive tract and act on the outside of the gut cell via receptors. Baculovirus recombinants express their *Bt* toxin inside the cell and the question is whether or not this internal toxin has insecticidal activity.

4) Upon infection ecdysone is not inactivated by glycosylation and hence a natural moult occurs.

5) Allatotropin promotes the release of JH and the maintenance of the juvenile stage of the insect. These insects continue to feed until they are affected by the virosis.

9.7

Two main concerns are:

• toxicity of the recombinant protein for non target organisms;

• gene transfer to the other organism.

9.8

1) Gram quantities. The OB content in diseased insects is about one third of their body weight.

2) A 2 litre bioreactor is needed. Cells grow at a density of 10^6 per ml. Therefore we would produce 5×10^7 OBs per ml (50×10^6). To produce 10^{11} we would require $10^{11} \div 5 \times 10^7$ ml $= 2 \times 10^3$ ml $= 2$ litres.

3) The non occluded form of the virus causes systemic infection in the insect and is present in the haemolymph. Cell culture or haemolymph derived NOVs can be administered by injection directly into the haemolymph, but this procedure is impractical on a large scale and requires considerable experimental skills.

4) In these sources many embryonic or frequently dividing cells are present. These have the ability to grow rapidly in tissue culture.

9.9

The inundative release of baculovirus is preferred, since the population densities are relatively low, the crops are frequently harvested removing virus from the site and damage cannot be tolerated. It is unlikely that the virus will persist on the site.

Responses to Chapter 10 SAQs

10.1   3 µl of sample was applied. To be detected we need 125 pg of viroid. Therefore 3 µl of sample must contain at least 125 pg of viroid. Therefore 1 µl must contain 125 divided by 3 pg (= 42 pg) of viroid. Therefore 0.5 ml must contain 42 x 500 pg of viroid = 21 000 pg = 21 ng. But the 0.5 ml sample was produced from 1g of leaf tissue. Therefore the system will enable detection of 21 ng of viroid in 1g of leaf.

10.2

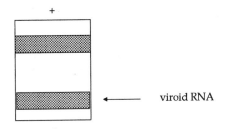

After the first electrophoresis, we end up with a mixture of RNAs at the end of the gel. Heating denatures the largely internally based paired RNA of the viroid. The second electrophoresis is run in the opposite direction. In this case the 'opened', now single stranded, viroid RNA migrates slower and so lags behind the rest of the RNAs. The band closest to the negative pole therefore represents viroid RNA.

10.3   Plant viruses can be detected in two main ways: either using methods directed against the coat protein or methods against the nucleic acids of the genome. Most detection methods used are of the ELISA type with antibodies against the coat protein. The reason is that this method is very sensitive, it can be used on a large variety of tissues and it can be easily automated. Furthermore ELISA methods are quick and relatively inexpensive. The molecular hybridisation technique has similar sensitivity but is difficult to automate and is time consuming and is, therefore, more costly to conduct.

Appendix 1

Units of measurement

For historical reasons a number of different units of measurement have evolved. The literature reflects these different systems. In the 1960s many international scientific bodies recommended the standardisation of names and symbols and a universally accepted set of units. These units, SI units (Systeme Internationale de Unites) were based on the definition of: metre (m), kilogram (kg); second (s); ampare (A); mole (mol) and candela (cd). Although, in the intervening period, these units have been widely adopted, their adoption has not been universal. This is especially true in the biological sciences.

It is, therefore, necessary to know both the SI units and the older systems and to be able to interconvert between both sets.

The BIOTOL series of texts predominantly uses SI units. However, in areas of activity where their use is not common, other units have been used. Tables 1 and 2 below provides some alternative methods of expressing various physical quantities. Table 3 provides prefixes which are commonly used.

Mass (S1 unit: kg)	Length (S1 unit: m)	Volume (S1 unit: m^3)	Energy (S1 unit: $J = kg\ m^2\ s^{-2}$)
$g = 10^{-3}\,kg$	$cm = 10^{-2}\,m$	$l = dm^3 = 10^{-3}\,m^3$	$cal = 4.184\ J$
$mg = 10^{-3}\,g = 10^{-6}\,kg$	$\text{Å} = 10^{-10}\,m$	$dl = 100\ ml = 100\ cm^3$	$erg = 10^{-7}\ J$
$\mu g = 10^{-6}\,g = 10^{-9}\,kg$	$nm = 10^{-9}\,m = 10\text{Å}$	$ml = cm^3 = 10^{-6}\,m^3$	$eV = 1.602 \times 10^{-19}\ J$
	$pm = 10^{-12}\,m = 10^{-2}\,\text{Å}$	$\mu l = 10^{-3}\,cm^3$	

Table 1 Units for physical quantities

Concentration (SI units: mol m^{-3})
a) $M = \text{mol l}^{-1} = \text{mol dm}^{-3} = 10^3 \text{ mol m}^{-3}$
b) $\text{mg l}^{-1} = \mu\text{g cm}^{-3} = \text{ppm} = 10^{-3} \text{ g dm}^{-3}$
c) $\mu\text{g g}^{-1} = \text{ppm} = 10^{-6} \text{ g g}^{-1}$
d) $\text{ng cm}^{-3} = 10^{-6} \text{ g dm}^{-3}$
e) $\text{ng dm}^{-3} = \text{pg cm}^{-3}$
f) $\text{pg g}^{-1} = \text{ppb} = 10^{-12} \text{ g g}^{-1}$
g) $\text{mg\%} = 10^{-2} \text{ g dm}^{-3}$
h) $\mu\text{g\%} = 10^{-5} \text{ g dm}^{-3}$

Table 2 Units for concentration

Fraction	Prefix	Symbol	Multiple	Prefix	Symbol
10^{-1}	deci	d	10	deka	da
10^{-2}	centi	c	10^2	hecto	h
10^{-3}	milli	m	10^3	kilo	k
10^{-6}	micro	μ	10^6	mega	M
10^{-9}	nano	n	10^9	giga	G
10^{-12}	pico	p	10^{12}	tera	T
10^{-15}	femto	f	10^{15}	peta	P
10^{-18}	atto	a	10^{18}	exa	E

Table 3 Prefixes for S1 units

Appendix 2

Chemical Nomenclature

Chemical nomenclature is quite a difficult issue especially in dealing with the complex chemicals of biological systems. To rigidly adhere to a strict systematic naming of compounds such as that of the International Union of Pure and Applied Chemistry (IUPAC) would lead to a cumbersome and overly complex text. BIOTOL has adopted a pragmatic approach by predominantly using the names or acronyms of chemicals most widely used in biologically-based activities. It is recognised however that there remains some potential for confusion amongst readers of different background. For example the simple structure CH_3COOH can be described as ethanoic acid or acetic acid depending on the environment or industry in which the compound is produced or used. To reduce such confusion, the BIOTOL series makes every effort to provide synonyms for compounds when they are first mentioned and to provide chemical structures where clarity and context demand.

Appendix 3

Abbreviations used for the common amino acids

Amino acid	Three-letter abbreviation	One-letter symbol
Alanine	Ala	A
Arginine	Arg	R
Asparagine	Asn	N
Aspartic acid	Asp	D
Asparagine or aspartic acid	Asx	B
Cysteine	Cys	C
Glutamine	Gln	Q
Glutamic acid	Glu	E
Glutamine or glutamic acid	Glx	Z
Glycine	Gly	G
Histidine	His	H
Isoleucine	Ile	I
Leucine	Leu	L
Lsyine	Lys	K
Methionine	Met	M
Phenylalanine	Phe	F
Proline	Pro	P
Serine	Ser	S
Threonine	Thr	T
Tryptophan	Trp	W
Tyrosine	Tyr	Y
Valine	Val	V

Index

A

acetolactate synthase, 212 , 213
acetosyringone, 191
advantages of molecular plant breeding, 185
agarose beads, 128
agglutination with PEG, 120
agricultural pressures, 51
Agrobacterium, 118 , 198
     host range, 191 , 193
Agrobacterium rhizogenes, 21 , 187 , 219
Agrobacterium tumefaciens, 21 , 34 , 187
agrochemicals, 12
agropine, 188
airlift reactor, 239
alfalfa mosaic virus, 203
allelic replacement, 232
allopolyploids, 41
Alstroemeria, 59
amidation, 231
amiprophos-methyl (APM), 133
amylose-free starch, 54
aneuploids, 83 , 102
anthers, 62
anti-herbicide lobby, 29
anti-trust laws, 31
antibody screening, 167
antisense, 206 , 218
antisense mRNA, 180
antisense technique, 218
antisense viral RNA, 207
antiviral proteins, 228
apical dominance, 187
apical meristem, 72
apidaecins, 215
apideacines, 23
Apis mellifera, 215
apple, 193
artificial T-DNA, 195
asparagus, 193
asymmetric hybrid plants, 132
asymmetric hybrids, 132
atrazines, 29
Atropa belladonna, 132
Autographa californica, 226
autopolyploid, 41
auxin, 187
auxotrophs, 112

B

B. thuringiensis, 24
     See also *Bacillus thuringiensis*
B. thuringiensis as a bioinsecticide, 208
Bacillus thuringiensis, 21 , 23 , 30 , 234
Bacillus thuringiensis crystal protein genes, 207
bacteria, 23
bactericidal peptides, 215
bactericidal properties, 23
baculovirus DNA, 232
baculoviruses, 222
     insecticides, 233
barley, 41 , 43
beet, 199
bi-phasic replication cycle, 228
bialophos, 213
binary systems, 195
Bintje, 13 , 33
bio-insecticides, 207
biological control of pests, 13
biotic factors, 40
border sequences, 191
Brassica, 132
Brassica napus, 133
Brassica oleracea, 53
breeders rights, 31
breeding companies, 30
breeding strategies, 13
bulbs, 59

C

CAT gene, 196
Catharanthus roseus, 254
cauliflower, 193
cauliflower mosaic virus, 197
cDNA, 9 , 170
cDNA library, 158
cDNA probes, 250
cecropins, 215
cell fusion, 5 , 12 , 75
cell selectable markers, 108
cell sorting, 127
cells, 6
certification of disease free status, 244
chalcone synthase, 218
chemical identification, 250
chemical pesticides, 222
chemotherapy, 72
chimera, 83
chimeric genes, 195
chimeric somatic embryos, 201
chitinase, 217

chloramphenicol, 196
chloramphenicol acetyltransferase (CAT), 196
chloroplast, 198
chloroplast markers, 130
chloroplasts, 129
chromosomal virulence, 189
chromosome elimination, 122 , 132
chromosome stability, 104
chromosome walking, 172
chrysanthemums, 193 , 248
chv genes, 194
      See also chromosomal virulence genes
chymotrypsin, 211
cis systems, 195
citrus, 203
Clavibacter michiganense pathovars, 216
clonal selection, 91
clone, 43
cloning, 4
cloning of cDNA, 160
co-infection, 236
CO_2 fixation, 198
colchicine treatment, 57
conservation of germplasm, 79
constitutive promoter (35S CaMV), 203
copy DNA, 9
      See also cDNA
cotton, 58 , 199
cpDNA, 53 , 58
crecropines, 23
crop protection, 12 , 40
cross-fertilisation, 60
cross-pollinators, 61
crossing, 47
crossing over, 138
crossing over frequency, 145
crown gall, 186
cruciferae, 133
cryopreservation, 79
crystal protein gene, 208
crystal protein gene transfer, 210
cucumber mosaic virus, 203
cultivar, 43
cultivars
      maintenance of, 48
culture at low cell density, 128
culture of heterokaryons, 128
cybridisation, 62
cybrids, 119
cytokinin, 198
cytological analysis, 129
cytoplasmic inheritance, 58
cytoplasmic male sterility, 53 , 61 , 130 , 132
cytoplasmic polyhedrosis viruses, 225

cytoplasmic strands, 126
cytoplasts, 125
cytotoxicity of PEG, 121

D

Daucus, 132
delayed early (DE) genes, 228
detection of bacteria, 250
detection of fungi, 252
detection of viruses, 248
diagnostics, 244
dicotyledonous plants, 193 , 200
dielectrophoresis, 122
differential screening, 175
differentially expressed genes, 176
digestive tract, 228
dihydroflavonol 4-reductase, 219
Dioscorea, 193
Diptera, 223
disease free, 66
diversification of metabolic products, 219
DNA
      methylation, 106
DNA sequence determination, 177
donor protoplasts, 119

E

earliness, 40
early genes, 230
ecdysone UDP glucosyl transferase (EGT), 230 , 236
ecdysteroids, 230
ecological hazard, 237
electrofusion, 122
electroporation, 118 , 200
ELISA, 248 , 250 , 251
embryo rescue, 4 , 80
embryos, 66
endonuclease activity, 191
endosperm, 215
enolpyruvylshikimate-3-phosphate synthase, 213
enrichment, 112
environmental effects, 18
environmental hazards, 222
environmental issues, 2
environmental variability (Ve), 49
epigenetic changes, 100 , 108
epithelial cells, 207
epizootic, 223
epizootiological potential, 237
Erwinia, 216
Erwinia chrysanthemi, 251 , 254

Erwinia spp., 70
erythrorhizon, 114
ethyl-N-nitrosourea, 109
ethylmethanesulphonate, 109
euploid, 83
evolutionary pressures, 51
exoskeleton, 228
exotic plants, 25
extra-chromosomal inheritance, 102
exudates, 29

F

F$_1$ generation, 49
F$_2$ generation, 49
fat body, 226
feeder cells, 128
feeder layer system, 111
fingerprint, 129
flavonoids, 218
flax, 199
flower colour, 218
fluorescein diacetate (FDA), 126
fluorescence activated cell sorter, 127
fruit softening, 219
fungi, 23
fungicides, 223
fusion chambers, 123

G

galactosidase, 232
gene amplification, 106 , 185
gene banks, 31
gene depletion, 106
gene isolation, 158 , 185
gene pool, 62
gene regulation, 228
gene transfer, 20
gene transfer experiments, 179
genes for transfer, 202
genetic analysis, 179
genetic engineering, 52
genetic erosion, 27
genetic manipulation, 62
genetic variance (Vg), 49
genetic variations, 100
genomic DNA, 170
Giemsa banding, 129
glucuronidase gene (GUS), 196
        See also GUS
gluten, 5
glycine rich protein, 217
glycosylation, 231

glyphosate, 213
Gramineae, 215
grasses, 193
growth retardants, 80
GUS, 196
GUS activity, 196
GUS reporter genes, 201

H

haemolymph, 226
hairy root disease, 187
haploid, 41 , 139
haploid production, 67
heat shock promoters, 197
herbicide resistance, 212
herbicides, 24 , 30
heritability, 48
heterokaryons, 118 , 126
heterologous probe, 170
heterozygous, 41
high energy radiation, 57
homozygous, 41
host range, 222
Hyalaphora cecropia, 215
hybrid variety, 60
hybrid vigour, 126
hybridisation, 129 , 154
hydroxyacetosyringone, 191
hygromycin, 196
hygromycin phosphotransferase (HPT), 196
hyperexpression, 229
hypersensitive response (HR), 216
hypocotyl tissue, 119

I

iaaH, 187
iaaM, 187
immediate early (IE) genes, 228
immunoblot analysis, 232
immunofluorescence, 232
immunofluorescence colony-staining, 251
in vitro recombination, 8 , 10
in vitro translation, 178
incompatability, 61
induction of tuberisation, 77
inhibition of gene expression , 24
insect cell cultures, 239
insect cell lines, 239
insect host range, 208
insect resistance, 207
insect species, 29
insects, 23

inter-specific hybridisation, 52
intracellular receptors, 205
introgression, 57 , 132
introns, 9
isoenzymes, 129

K

kanamycin resistance (NPTII), 210
killer genes, 19

L

lambda vectors, 161
latex-agglutination test, 249
Lepidoptera, 223
lettuce, 193 , 199
leucinopine, 188
light driven promoters, 197
line, 43
linkage, 138
Lithospermum, 114
Lolium perenne, 44
luciferase enzymes, 196
LUX genes, 196
Lycopersicon esculeutum, 56
Lycopersicon peruvianum, 56
lysine, 112

M

macro-injection, 201
maize, 54 , 193 , 200 , 201 , 219
mankind-curtailed plants, 19
mankind-dependent plants, 19
mankind-independent plants, 19
mannopine, 188
marker genes, 195
Mendelian inheritance, 49
Mendelian rules, 184
meristematic tissue, 193
mesophyll protoplasts, 126
methionin, 112
methyl-N-nitro-N-nitrosoguandidine, 109
micro-injection, 200
micro-injection of DNA, 186
microculture, 126
microculture chamber, 125
microfusion, 125 , 126
micromanipulator, 126
micropippette, 126
microprotoplasts, 133
microspores, 200
mitochondria, 129 , 198

mitochondrial markers, 130
mitotic crossing over, 106 , 199
modern varieties, 51
molecular breeding, 13
molecular hybridisation, 246 , 250 , 251
molecular markers, 5
molecular plant breeding, 185
moncogenic traits, 12
monoclonal antibodies, 245 , 248 , 253
monocotyledonous plants, 201
monogenic traits, 40 , 202
monogerm seeds, 40
monopoly positions, 30
morphological markers, 126
mtDNA, 53 , 58 , 62
multigenic traits, 13 , 14
mutagenesis, 103
mutagens, 109
mutants, 100
  dominant, 113
  recessive, 113
mutation, 51 , 52
mutation spectrum, 103
mutations, 102
  addition, 52
  deletion, 52
  recessive, 109
  replacement, 52

N

naked RNA, 245
natural plant resistance genes, 207
necrosis, 216
nematode resistant genes, 24
nematodes, 24
neomycin phosphotransferase (NPT), 195
new flower colours, 24
new genes and risk assessment, 21
new strain, 4
new viral pathogens, 222
nick translation, 246
Nicotiana, 133
Nicotiana plubaginifolia, 132
Nicotiana plumbaginifolia, 132
Nicotiana tabacum, 132
non-occlued virus particles = NOV, 226
nopaline, 188
NOV replication, 230
NPTII fusion, 210
nuclear polyhedrosis virus, 226
nucleo-organelle combinations, 129
nucleocapsids, 225

O

OB envelope gene, 230
OB morphogenesis, 229
occlusion bodies, 225
octopine, 188
oligonucleotide probes, 165
oncogenes (onc), 187
onion, 53 , 201
operons, 191
opines, 187 , 188
organ specific promoters, 197
organogenesis, 82
ortet, 4
    definition, 70
oryzalin, 133

P

p10 gene, 237
p10 promoters, 236
particle gun, 201
patent law, 31
pathogen eradication, 70
pathogenesis related (PR) proteins, 217
pesticides, 30 , 223
Petunia, 133 , 193
Petunia hybrida, 151
pharmaceuticals, 12
phenotypic variance, 49
phenylalanine, 218
phosphinothricine, 196 , 213
phosphinothricine acetyltransferase
(PAT), 213
phosphinothricine acetyltransferase gene,
(PAT), 196
phosphonitricine, 29
phosphorylation, 231
physiological response, 100
phytosanitary, 19
plant breeding
    conventional, 38 , 52
    conventional programmes, 4
    objectives, 3
    traditional, 2
plant pathogens
    principles for the detection of, 245
plant protection, 244
plant regeneration, 102
plant regeneration from hairy roots, 187
plant storage organs, 211
plantlets, 66
plasmids, 186
plasmolysation, 85

ploidy level, 118
poly-L-lysine, 200
poly-L-ornithine, 200
polyacrylamide-gel electrophoresis, 246
    bi-directional, 247
polycations, 120
polyclonal antibodies, 245
polyethylene glycol (PEG), 120
polygalacturonase, 219
polygenic traits, 40 , 202
polyhedrin, 229 , 230
polyhedrin gene, 231 , 238
polyhedrin promoters, 236
polymerase chain reaction, 168 , 253
polyploid, 41
polyvinylpyrrolidone, 252
poplar, 199 , 210
post translational modifications, 231
potato, 33 , 59 , 67 , 69 , 100 , 133 , 193 , 199 , 203 ,
210 , 214 , 219 , 248
potato inhibitor II, 211
potato virus X (PVX), 203
probes, 5
proctolin, 234
promoter activities, 185
promoters, 185 , 195 , 197
    2', 197
    35S, 197
proteinase inhibitors, 23 , 24 , 30 , 211
protoplast fusion, 62 , 119
protoplasts, 14 , 62 , 85 , 119 , 125
protoxin, 208
Pseudomonas, 216
public debate, 27
purified DNA, 200

Q

quality, 40

R

ramet
    definition, 70
ramets, 4
rapeseed, 58 , 193 , 199 , 201 , 210 , 214
Raphanus sativum, 58
rapid multiplication, 72
recessive, 43
recessive mutations, 102
recombinant DNA technology, 8 , 52 , 75
red chlorophyll fluorescence, 126
reducing virus spread, 237
regeneration, 108

regulation signals, 185
release of baculoviruses, 239
reporter genes, 195 , 196
restricition fragment length polymorphism, 152
restriction fragment length polymorphism, 129
    See also RFLP
reverse-transcriptase, 254
RFLP analysis, 129
RFLP linkage maps, 129
Rhizobium spp., 21
Rhizoctonia solani, 70
rhizomes, 59
R_i (root inducing) plasmids, 187
rice, 54 , 193 , 200 , 201
risk assessments, 29
rol genes, 187
rolA, 187
rolB, 187
rolC, 187
rolD, 187
rose, 193
Russel Burbank, 13
rust resistance, 58
rye, 44

S

S/T ratio, 250
Salmonella typhimurium, 213
satellites, 205
screening, 232
secondary embryos, 201
secondary metabolites, 88
secondary products, 12
segregation, 105 , 138
selectable gene, 8
selectable marker genes, 196
selectable markers, 167
selection, 48 , 84 , 108 , 126
selection by biochemical markers, 128
selection of hybrids, 86
selective media, 250
self-fertilisation, 59
self-fertilising crops, 61
serology, 248
sex factors, 186
sexual crossing, 103
sexual hybridisations, 129
sexual hybrids, 118
shelf life, 219
shoot cultures, 77
shoot meristems, 201
single stranded T-DNA, 191
skimmed milk powder, 252

Solanaceae, 133
Solanum dulcamara, 33
Solanum nigrum, 33
Solanum tuberosum, 33 , 68 , 89 , 132 , 133
somaclonal variation, 102
somatic embryogenesis, 67 , 81
somatic hybridisation, 52 , 85 , 134 , 184
somatic hybrids, 118 , 126
somoclonal variation, 103
Southern blot, 232
Southern blotting, 154
species extinction, 29
specificity of antibodies, 249
specificity of baculoviruses, 223 , 237
spindle toxins, 133
Spiroplasma citri, 254
Spodoptera frugiperda, 226 , 239
starch, 219
stereo-specific products, 88
sterile plants, 184
sterility
    induced, 19
Streptomyces hygroscopicus, 213
sugar beet, 210
sulphonyl urea compounds, 212
sunflower, 58 , 193 , 199
suppressed replication, 206
surface sterilisation, 76
survival curve, 109

T

T-DNA, 187 , 194
T-DNA 2' promoters, 210
T-DNA border sequences, 194
T-DNA transfer, 191
T-DNA transfer system, 194
tandem DNA copies, 205
targeting gene products, 198
telotrisomics, 151
terminators, 195
thermotherapy, 71
thionins, 23 , 215
third world markets, 32
three point test crosses, 149
T_i plasmid, 189
tobacco, 193 , 199 , 203 , 210
tobacco mosaic virus (TMV), 203 , 217
tobacco ringspot virus (TobRV), 205
tobacco spotted virus (TSV), 203
tomato, 193 , 199 , 203 , 210 , 214 , 219
tomato cultivar, 246
totipotent plant cells, 198
tracheal matrix, 226

trans systems, 195
transactivators, 230
transducing phages, 186
transduction, 186
transformation, 186
transformation efficiency, 200
transformed cells, 199
transgenic plant cells, 185
transgenic plants, 18
transgenic rye plants, 201
transgenic tobacco plants, 201 , 210
translocation, 57
transposable elements, 105
transposon, 174
transposon tagging, 174
triazines, 113
Trichoplusia, 239
triploid, 41
trisomics, 151
trypsin, 211
trypsin inhibitor, 234
tuberisation, 77
tubers, 59
tulips, 59
tumour inducing plasmids, 187
tumour specific metabolites, 188
two point crosses, 142

U

upstream TATA box, 230

V

variability, 49
vector, 118
vectors, 163 , 231
vegetative propagation, 59
Vibrio harveyi, 196
vir (virulence) genes, 189
vir genes, 194
        See also virulence genes
virA protein, 191
viral coat protein, 22
viral coat proteins, 30
viral RNA molecules, 205
virC proteins, 191
virD proteins, 191
virD1, 191
virD2, 191
virD2 protein, 192
viroids, 70 , 245
        detection, 245
virus particles, 225

virus resistance, 203
virus resistant plants, 203
viruses, 20 , 22 , 70 , 186
visualisation of hetrofusions, 120

W

weeds, 18
wheat, 5 , 41 , 43 , 58 , 193 , 200
wild types, 102
wound callus, 193

X

Xanthamonas campestris, 216

Z

zygotes, 41